Feminist Technosciences

Rebecca Herzig and Banu Subramaniam, Series Editors

Molecular Feminisms

BIOLOGY, BECOMINGS, AND LIFE IN THE LAB

DEBOLEENA ROY

UNIVERSITY OF WASHINGTON PRESS
Seattle

Publication of this open monograph was the result of Emory University's participation in TOME (Toward an Open Monograph Ecosystem), a collaboration of the Association of American Universities, the Association of University Presses, and the Association of Research Libraries. TOME aims to expand the reach of long-form humanities and social science scholarship including digital scholarship. Additionally, the program looks to ensure the sustainability of university press monograph publishing by supporting the highest quality scholarship and promoting a new ecology of scholarly publishing in which authors' institutions bear the publication costs.

Funding from Emory University and the Andrew W. Mellon Foundation made it possible to open this publication to the world.

www.openmonographs.org

Copyright © 2018 by Deboleena Roy
Printed and bound in the United States of America
Interior design by Thomas Eykemans
Composed in Chaparral, typeface designed by Carol Twombly
Cover design by Katrina Noble
Cover photograph by Kheyal Roy-Meighoo and Koan Roy-Meighoo

22 21 20 19 5 4 3 2

All rights reserved. No part of this publication may be reproduced or transmitted in any form or by any means, electronic or mechanical, including photocopy, recording, or any information storage or retrieval system, without permission in writing from the publisher.

UNIVERSITY OF WASHINGTON PRESS
www.washington.edu/uwpress

LIBRARY OF CONGRESS CATALOGING-IN-PUBLICATION DATA
LC record available at https://lccn.loc.gov/2018010432

ISBN 978-0-295-74409-4 (hardcover), ISBN 978-0-295-74410-0 (pbk), ISBN 978-0-295-74411-7 (ebook)

The text of this book is licensed under a Creative Commons Attribution-Non Commercial-NoDerivatives 4.0 International License (CC BY-NC-ND 4.0): https://creativecommons.org/licenses/by-nc-nd/4.0.

To my parents, Ila and Manick Roy, for inspiring me to work hard

To my sister, Madhumeeta Roy, for encouraging me to dream big

To my partner, Sean Meighoo, and our children, Koan and Kheyal Roy-Meighoo, for reminding me to laugh and play every day

Contents

Acknowledgments

I have been torn between the natural sciences and humanities for a long time. In fact, this book has probably been in the works since the day I sent in my acceptance to pursue a doctoral degree in molecular biology and reproductive neuroendocrinology. By accepting that admissions offer and extremely rewarding academic path, I declined an offer from an environmental studies program where my plan was to study bioethics, environmental justice, and women's reproductive health movements. At my side while I made this difficult decision about my academic trajectory was a cadre of feminists who encouraged me to stay in the "hard" sciences, to become a feminist scientist, to learn the science that was being used to develop new reproductive and genetic technologies, and to participate in the creation of scientific knowledge. A long time coming, this book is a direct result of that encouragement.

At any given point in my intellectual journey, I have been unbelievably fortunate to have many generous feminist mentors at my side. Several of them welcomed me into their midst while I was just an undergraduate student studying microbiology. Laura Sky introduced me to some of the most fierce and eloquent scholar-activists I have ever come across, including Shree Mulay, Karen Messing, Sunera Thobani, and Elizabeth Abergel. As a master's student studying in a cancer biology lab at McMaster University, my supervisor Andrew Rainbow supported my interdisciplinary tendencies; when I had finished required course work in genetics and radiobiology, he encouraged me to take a directed reading course in the philosophy department. For this opportunity, I am forever grateful. Through this course, Elizabeth Boetzkes introduced me to the work of Sandra Harding, Helen Longino, Nancy Tuana, Alison Wylie, and Joseph

Rouse. Reading their work as a graduate student gave me a sense of what becoming a feminist scientist would actually entail. I never imagined that one day I would have the opportunity to meet and share my ideas with these feminist philosophers in person.

As a doctoral student, I had the chance to study with one of the very few female professors in the Department of Physiology at the University of Toronto. I am thankful to my supervisor, Denise D. Belsham, for sharing her scientific curiosity and her love of molecular biology research with me. I am grateful to Neil MacLusky, Ted Brown, and Bernardo Yusta for their guidance and support during my doctoral research. I want to acknowledge my lab mates Arawn Therrian, Tarranum Shakil, and Monica Antenos for their friendship and support. I also want to acknowledge the many nonhuman actants in the lab for their role in teaching me about molecular biology. It was, however, the course I took with Margrit Eichler in medical sociology and reproductive technologies at the University of Toronto that led me to my interdisciplinary career path. I can never thank Margrit enough for allowing me to audit her graduate seminar, introducing me to the field of feminist science studies, requiring me to write a final paper for the course, and encouraging me to submit that paper to the journal *Hypatia* while I was still a graduate student. I was so fortunate to receive this mentorship as a young scholar, and I am inspired to try to do the same for my own students. Through Margrit I was invited to join the Social Sciences and Humanities Research Council–funded group Biology As If the World Matters (BAITWorM), where Linda Muzzin and the late Peggy Tripp-Knowles showed me that social justice work and scientific research can go hand in hand.

This brief exposure to the field of feminist science studies and my participation in a science and social justice organization made me take a gamble and apply for a tenure-track position in a women's studies department. What reason my colleagues in the Women's Studies Department at San Diego State had to take a gamble on me, I will never know. I am forever grateful to Huma Ahmed-Ghosh, Irene Lara, Susan Cayleff, Pat Huckle, Anne Donadey, Doreen Mattingly, Oliva Espin, Esther Rothblum, Bonnie Zimmerman, and Bonnie Kime Scott for supporting me while I trained to become a women's studies scholar. Around this time I was invited into an incredibly supportive community of feminist science and technology studies (STS) scholars. Two particular events during this period stand out.

First, while giving a talk at the inaugural Feminist Epistemologies, Methodologies, Metaphysics and Science Studies (FEMMSS) conference in Seattle, I had the good fortune of meeting Sandra Harding and Karen Barad. Both of these influential feminist thinkers asked me extremely difficult questions after my talk; to this day, they continue to generously engage in and support my work. Second, I was invited to join the Next Wave in Feminist Technoscience group by Jennifer Terry and Kavita Philip at the University of California–Irvine, where I was introduced to Banu Subramaniam and Michelle Murphy, both of whom became thoughtful interlocutors. Thanks to these mentors, I developed a vocabulary in feminist STS.

During my time at Emory University, I have been able to carry out several interdisciplinary projects. It is rare to have the support of an institution that doesn't ask me to justify why I want to bring the humanities and the sciences together, or waste my energy and time making the case that women's studies and the neurosciences should come together. Instead, I get to spend my time doing what I love. Thanks to my joint appointment in the Department of Women's, Gender, and Sexuality Studies (WGSS) and in the Program of Neuroscience and Behavioral Biology (NBB), I teach women's studies students about molecular biology and neuroscience, and neuroscience students about feminism. I join communities of learners across campus who appreciate the importance of interdisciplinary pursuits. My colleagues in the Department of WGSS have been unwavering in their support. I thank Lynne Huffer, Pamela Scully, and Elizabeth Wilson for being fearless chairs and generous mentors, and for creating a department culture where academic curiosity and friendship go hand in hand. I also want to acknowledge my WGSS colleagues Carla Freeman, Michael Moon, Irene Browne, Beth Reinghold, Falguni Sheth, Kadji Amin, and Calvin Warren for their support.

To my colleagues in NBB—namely Paul Lennard, Kristen Frenzel, Michael Crutcher, Keith Easterling, Bob Wyttenback, Leah Anderson Roesch, and Gillian Hue—I thank you for your openness to my interdisciplinary wanderings and for supporting feminist STS at the institutional and curricular levels. The administrative staff in both of my academic homes solve problems for me every day, and for this I am truly grateful. They are Neema Oliver, Marybeth Smith, Kashira Baker, Alan Weinstein, and Nadia Brown-Ware. To my many colleagues in the Graduate Division of Biological and Biomedical Sciences, I thank you for the opportunity to actually practice the art of having difficult conversations and developing shared moments of perplexity. In particular, I acknowledge my colleagues Donna Maney, Ron Calabrese, Shawn Hochman, Yoland Smith, and David Weinshenker in the Neuroscience Program, for supporting my neuroscience research and teaching efforts. I also thank Ichiro Matsumura in the Department of Biochemistry for the lively discussions on synthetic biology, ethics, and life in the lab.

Paul Root Wolpe, Karen Rommelfanger, and Arri Eisen at Emory's Center for Ethics have supported my research and teaching during my time as a senior fellow. Many more incredible faculty colleagues, working across several different disciplines, have supported my work and extended their friendship, including Yanna Yannakakis, Ben Reiss, Rosemarie Garland-Thomson, Aiden Downey, Deepika Bahari, Jonathan Goldberg, Sander Gilman, Stu Marvel, and Cindy Willett. I also thank the students that I have had the great fortune of working with at Emory. In particular, I thank students from my recent WGSS graduate seminars in feminist and postcolonial STS, and my seminar on feminism, Deleuze, and biology. They are Stephanie Alvarado, Jordan Johnson, Lamo Lamaocuo, Allison Pilatsky, Ingrid Meintjes, Samantha VanHorn, Abidemi Fasanmi, Katherine Bryant, Anna Kurowicka, Caroline Warren, Samia Vasa, Lily Oster, Kevin McPherson, and Stephanie Koziej. I also thank my graduate research assistant Dayne Alexander for kindly making sense of unruly lists of keywords and managing to find journal articles to help with my research.

I have received several fellowships and grants over the years. These scholarly gifts have provided me with much needed time to think, research, and write. They include a visiting scholar appointment at the Pembroke Center for Teaching and Research on Women at Brown University. I thank Londa Schiebinger and the Clayman Institute for Gender Research at Stanford University for supporting my research through a faculty fellowship. The actual serious thinking, research, and writing for this book project began, however, when I received a faculty fellowship from the Bill and Carol Fox Center for Humanistic Inquiry at Emory. I thank Tina Brownley, Keith Anthony, Amy Erbil, and Colette Barlow for supporting me during that magical year with stimulating conversations and an office fridge that was always full of goodies.

I want to acknowledge the National Academies KECK Futures Initiative for a grant that supported my research on synthetic biology and ethics. I also want to acknowledge the National Science Foundation for my Scholars Award. This book would only be an idea accompanied by some scribbled notes were it not for this award. In particular, I thank my program officer, Wenda Bauchspies, for her support. Thanks also to two graduate research assistants, Ingrid Meintjes and Kristie Garza, who brought their own excitement and expertise to this project. This book contains the theoretical framework I developed to carry out the research for my NSF award. The actual data and results of that NSF project will have to wait for the next book. Segments of this book draw from previously published works. A portion of chapter 2 appeared as the chapter "Feminist Approaches to Inquiry in the Natural Sciences: Practices for the Lab" in the *Handbook of Feminist Research: Theory and Praxis* (Sage Publications, 2011). Chapter 3 includes an earlier version of work that appeared as the chapter "Science Studies" in *The Oxford Handbook of Feminist Theory* (Oxford University Press, 2016). An earlier version of chapter 4 appeared as "Should Feminists Clone? And If So, How? Notes from an Implicated Modest Witness" in the Spring 2008 issue of the journal *Australian Feminist Studies*.

In addition, segments of this book have been presented by way of several invited talks while still in preparation. For the opportunity to share my work, I am thankful to the following: the Compassionate Knowledge Working Group at Hampshire University, the Gender Research Institute at Dartmouth University, the Program on Science and Technology Studies at Harvard University, the Neuroscience Lecture Series at Georgia State University, the Center for the Humanities at Wesleyan University, the Institute of Women's Studies and Gender Research at Johannes Kepler University, and the Life Un(Ltd) Lecture Series at the University of California at Los Angeles.

I have looked forward to writing these acknowledgments for some time. Even when it was unclear where the chapters of this book would take me, the support that I received from so many friends was always clear. To my STS, feminist STS, and feminist philosophy of science friends, I thank you for the stimulating discussions and for inspiring me to become a better thinker. You are Jenny Reardon, Anne Pollock, Jennifer Singh, Nassim Jafarinaimi, Jennifer Hamilton, Carol McCann, Ruha Benjamin, Rachel

Lee, Aimee Bahng, Kalindi Vora, Neda Atanasoski, Angie Willey, Sara Giordano, Moya Bailey, Catherine Hundleby, Alexis Shotwell, and Sharyn Clough. To my feminist and postcolonial STS working group friends, I thank you for the many nourishing conversations. You are Laura Foster, Kim TallBear, Sandra Harding, and Banu Subramaniam. To my friends in the NeuroGenderings Network who are both feminist neuroscientists and scientist feminists, you have taught me how to ask difficult questions, but more important you have shown me how to have fun while we try to answer these questions together. You are Anelis Kaiser, Isabelle Dussage, Sigrid Schmitz, Catherine Vidal, Cynthia Kraus, Rebecca Jordan-Young, Cordelia Fine, Emily Ngubia Kessé, Victoria Pitts-Taylor, Daphna Joel, Gina Rippon, Sari van Anders, Gillian Einstein, Cordelia Fine, Robyn Bluhm, Hannah Fitsch, Giordana Grossi, and Christel Gumy.

At the University of Washington Press, special thanks go to my extremely patient series editors, Banu Subramaniam and Rebecca Herzig, and my editor, Larin McLaughlin. Thank you so much for believing in this book even before I did. Also, I would not have been able to bring this book to its final form without the help of my copyeditor Amy Smith Bell and Beth Fuget and Julie Van Pelt at UW Press. For providing insightful and generous comments, I thank the two anonymous readers. I also want to specially thank a new friend, Sushmita Chatterjee, who falls into a category of her own. Despite only having known me for a month, Sushmita not only agreed to read my entire manuscript but also gave me an idea for the conclusion. She joins the many other feminists who are committed to the political work of academic community building. I am lucky to call you my friends.

I must end by acknowledging my family for their encouragement as I wrote every sentence in this book. To my partner's parents, Cynthia and Peter Meighoo, thank you for asking me about my work on a regular basis and for supporting my scholarly pursuits. Your generosity and kindness has helped me balance having a career and a family. To my own parents, Ila and Manick Roy, thank you for encouraging me to become a scientist from a young age. Yet, all the while you did this, you both also exposed me to the worlds of poetry, music, and dance. As a child of Bengali immigrants in Canada, I learned to sing an endless number of songs from Rabindranath Tagore's *Gitabitan* (1960), and as I did, my mother would carefully translate the meaning of each song from Bengali to English for me. My favorite Tagore songs were those that turned to the lives of birds, cows, forests, flowers, and clouds to glean some of life's most important lessons on love, desire, devotion, and beauty. I'm not blaming my parents, but they might be responsible for my constant meanderings between the sciences and humanities. To my late brother, Atanu Roy, losing you when we did taught me to find meaning in my work and to treasure the people around me. Thank you for these hard life lessons. To my sister, Madhumeeta Roy, thank you for figuring out the importance of following your passions from a very young age. I have learned by your example. Thank you for being my partner in perseverance. Your role in seeing this book through, by way of constant encouragement, is worthy of coauthorship. To my children, Kheyal and Koan Roy-Meighoo, your love sustains me and your creativity inspires me. Watching you grow has been a humbling experience, and I am perpetually in awe of how you encounter the world around you through curiosity and joy. In fact, as I write these acknowledgments, The Lollipops are jamming in the room beside me. Your version of "Heads Will Roll" by the Yeah Yeah Yeahs is nothing short of fierce.

Lastly, to my partner, Sean Meighoo, thank you for making time to read and comment on this manuscript and making it the first item of business on your to-do list during your post-tenure leave. Thank you for being a feminist, for introducing me to the scholarly field of women's studies, and for supporting me in my own feminist pursuits. Thank you most of all for putting together an awesome soundtrack for our lives together.

Molecular Feminisms

Introduction

Stolonic Strategies

> To be present at the dawn of the world. Such is the link between imperceptibility, indiscernibility, and impersonality—the three virtues. To reduce oneself to an abstract line, a trait, in order to find one's zone of indiscernibility with other traits, and in this way enter the haecceity and impersonality of the creator. One is then like grass: one has made the world, everybody/everything, into a becoming, because one has made a necessarily communicating world, because one has suppressed in oneself everything that prevents us from slipping between things and growing in the midst of things. One has combined "everything" (le "tout"): the indefinite article, the infinitive-becoming, and the proper name to which one is reduced.
>
> —GILLES DELEUZE AND FÉLIX GUATTARI

> stolon (runner): (i). A modified aboveground stem creeping and rooting at the nodes. (ii). It is an aerial shoot from a plant with the ability to produce adventitious roots and new clones of the same plant. Such plants are called stoloniferous. A stolon is a plant propagation strategy akin to a rhizome.
>
> —DINESH KUMAR AND YASHBIR SINGH SHIVAY

A few years back, a severe rainstorm brought down a mighty oak tree in my neighbor's backyard. The tree fell toward my house, with the top branches just scraping the roof above the bedroom where I slept. I've had several tree encounters in my life—from climbing trees as a child, to planting saplings as a girl scout, to walking through the mighty trunk of an ancient redwood in California. This particular tree encounter, however, was the first one that nearly ended my life. Despite this near-death encounter, I would not say that I am over trees, or even "tired of trees" as Gilles Deleuze and Félix Guattari's statement is often understood.[1] I've been around long enough to know that trees, like humans, have to live and die too. Interestingly, the event that was this tree spurred another very curious event that taught me a great deal.

The day the tree fell, I was in shock. This shock turned into immense awe as I watched countless animals and insects scurry in and out of the tree's topmost branches, which were now at my eye level. Had the tree trunk not been in imperceptible relationships with several different species of ants over a sixty- to seventy-year period, it would not have been hollowed by rot, too weak to withstand the wind gust that brought it down that day. It is not often that one has the opportunity to get up close to the leaves, branches, insects, and other tree becomings that generally transpire five to ten stories above ground. As much as they were opposed to arboreal thought, Deleuze and Guattari were also fully aware of such tree becomings. They stated: "A new rhizome may form in the heart of a tree, the hollow of a root, the crook of a branch. Or else it is a microscopic element of the root-tree, a radicle, that gets rhizome production going."[2] Indeed, by way of this fallen tree, I was brought into a new intimate relationship with grass, made keenly aware of the strategies used by grass stolons to grow and remake the world.

Removing the massive tree trunk and enormous branches from my backyard took several days, but once these were gone, I realized that practically all of the grass that had been growing, where the tree had fallen, was also gone. This tree may have deterritorialized when it was cut into hundreds of small pieces and removed from the yard, but the microscopic element of its root-tree made its way known through the remaining blades of grass that lay crushed upon my lawn. The easy answer would have been to lay down new sod. For several reasons (one of which was most certainly financial) I didn't want to replace the grass that had been destroyed, at least not immediately. I had a sense that the ground should have some time to recover from the blow, but more important I did not want to erase the event that was that tree so quickly.

By opening myself up to the route of least interference, I witnessed, over three full years, the slow processes of stolonic growth in a species of everyday backyard Bermuda grass. I literally spent time watching grass grow. I witnessed, as Dinesh Kumar and Yashbir Singh Shivay explain in their definition of a stolon, how creeping grasses spread not by a rhizomatic root system that is underground and generally invisible to human senses but rather by the stolonic processes of developing new shoots and extending horizontal stems that grow above ground.[3] Over time, I became captivated with the outwardly stretching veins that ran along the surface of the ground, constantly reaching out, in search of connections, feeling around. This is how I realized that grass has a strategy that works. This strategy is one of becoming, and as Deleuze and Guattari write, this strategy works at making a communicating world.

I use a similar strategy throughout this book to think more carefully about new connections and communications that can emerge between molecular biology and feminism. My hope is to show that by thinking with *molecular feminisms*, *biophilosophies of becoming*, and *microphysiologies of desire*, we can see that biology and biological processes need not be essentializing or deterministic. As a feminist scientist trained in molecular biology and reproductive neuroendocrinology, I often wonder what knowledges we as feminists might have created by now if we weren't constantly having to spend our time and energy producing counterclaims to essentializing or deterministic language, paradigms, and experimental designs in the biological sciences. What if we were able to use, as Audre Lorde suggested years ago, our "power of the erotic" to think about science, biology, and molecular biology?[4] By the term "erotic," Lorde was referring to the potential, desires, and creative forces that lie within us to create change. Angie Willey has recently drawn from Lorde's work to introduce the idea of "biopossibilities" and to encourage us to think differently about bodies and biologies.[5]

Can biopossibilities and the playfulness that comes with powers of the erotic change how we approach bodies and matter in the lab? Would feminists be more willing to "do biology" if we knew from the start that the outcomes of our research would not bind us to an unwelcome fate? What if, after all, anatomy was not our entire destiny but was indeed involved in the emergent and expressive processes of life forms in all their becomings? What if biological reductionism was not seen as an end to scientific knowledge but instead as a means to connect more intimately to the multiple microscopic and molecular material actants that make up the world within and around us? What if learning how to *see* the world was also about learning how to *encounter* that world? By using strategies inspired by grass, and by extending stolonic runners between new and old feminist engagements with science, I hope we will be able to think anew. What is at stake here is relevant to scholars in both the biological sciences and the humanities. It is the chance to learn how to approach bodies and matter through new lines of flight and to embrace the erotic possibilities and capacities that come with becoming a blade of grass.

Early in my scientific training, I conducted a series of lab experiments that inspired me to think about questions that lie smack-dab at the intersections between feminism, biology, and philosophy. I did my doctoral research in a molecular biology and reproductive neuroendocrinology lab that studies the molecular mechanisms of steroid regulation in an *in vitro* cell line of specialized hypothalamic neurons.[6] The goal of my research project was to further characterize these hypothalamic gonadotropin-releasing hormone (GnRH) neurons to better understand their role in the regulation of reproduction. One of the first experiments I contributed to involved applying molecular biology techniques to search for estrogen receptor gene expression and protein synthesis in these neurons.[7] Before I share more on this particular experiment, however, I want to point out the importance of learning the everyday nitty-gritty practices of experimentation in molecular biology and the impact that these practices have had on my development as a feminist scientist.

Throughout this book I emphasize the importance of developing practice-oriented approaches for feminist science and technology studies (STS). Perhaps driven by their background and training in the sciences, such feminist scholars as Donna Haraway, Isabelle Stengers, and Karen Barad have long highlighted the importance of these approaches.[8] Through her concept of cosmopolitics, for instance, Stengers has suggested that an emphasis on practices allows us to learn how to engage with and not simply judge a knowledge system that is not our own.[9] This shift from focusing on the construction of theories in science, to developing a better appreciation for experimental practices in science that can contribute to theory-making, is now a cornerstone of STS scholarship, perhaps best articulated by Ian Hacking, who has suggested that "experimentation has a life of its own."[10] I have tried to capture the erotic potential that lies within the life of experimentation, and by participating in this life I have gained a sense of what Deleuze and Guattari meant by reducing oneself to an abstract line and finding a zone of indiscernibility with other traits. Discussing the research techniques of protein modelers, Natasha Myers has recently stated that "modelers, it turns out, cultivate intimate relationships with their molecules as they get themselves caught up in the involving work of molecular visualization."[11] In my case, training in a molecular biology lab has led me to create intimate relationships and zones of indiscernibility not only with molecules but also with many other traits, including feminist activism, feminist theory, reproductive justice movements, neurons, genes, steroid receptors, steroids, signaling pathways, and philosophies of becoming.

The Life of an Experiment

At the beginning of my doctoral work, I had the opportunity to find new forms of communication between GnRH neurons, estrogen receptors (ERs), and estrogen molecules. The molecular biology techniques I was required to learn in order to conduct my experiments taught me how to face the lab bench, how to form experimental togetherness around shared objects of perplexity, and how to "break bread" as it were with my scientific peers and colleagues in the lab.[12] I longed to create new zones of proximity between molecules and my own feminist political landscapes by examining a biological interaction and process that had been marginalized in the sciences but was relevant to both molecular neuroendocrinology and women's reproductive health. Thus, my excitement when one of my first experimental research tasks involved searching for the expression of estrogen receptor genes and the synthesis of estrogen receptor alpha (ERα) and estrogen receptor beta (ERβ) proteins in an *in vitro* cell line model of GnRH neurons called GTI-7 cells.[13] GnRH is known to be a central hormone of the hypothalamic-pituitary-gonadal (HPG) axis. It helps to regulate the synthesis and secretion of luteinizing hormone and follicle stimulating hormone in the pituitary gland and androgen and estrogen in the gonads.[14] I was tasked with investigating the possibility for direct feedback regulation of GnRH gene expression and synthesis by the gonadal hormone estrogen.

Estrogen of course is involved in much more than just the processes of reproduction, including the fusion of long bone epiphyses, the suppression of osteoclast activity, and providing direct protective effects against atherosclerosis.[15] Although estrogens are synthesized primarily in the gonads, they are also synthesized in extra-gonadal sites and are known to play an important role in coordinating regulatory effects in adipose tissue, skin, the immune system, and more.[16] My doctoral supervisor at the University of Toronto, Denise D. Belsham, was interested primarily in examining the role of estrogen in the brain in its reproductive capacity. Estrogen is a very interesting hormone in this respect, as it has been shown to negatively regulate GnRH synthesis but is also necessary to induce the preovulatory surge of GnRH during the menstrual cycle.[17] One would imagine therefore that estrogen would have something to do with the regulation of reproduction at the level of the brain. The research we conducted would almost be considered unnecessary if it were not for the fact that experts in the field had for decades declared that GnRH neurons did not express the proper nuclear protein receptors that bind to estrogen. Thus the working premise in the field, and the belief held by most neuroendocrinologists at the time, was that GnRH neurons could not be directly affected by estrogen. Instead, interneurons contacting GnRH neurons were thought to be responsible for mediating the effects of estrogen and other gonadal steroids on GnRH synthesis and secretion.[18]

This is what I was up against when I started my doctoral research. I was working athwart to expert knowledge in reproductive neuroendocrinology that was unconvinced of a trait (that of a relationship between estrogen receptors and GnRH neurons) and unwilling or uninterested in spending time designing experiments to recognize new zones of indiscernibility. They dismissed this possibility because they could not imagine what estrogen would be doing in this part of the brain. The long-held belief that GnRH neurons functioned without experiencing direct contact by estrogen had resulted in a scientific milieu where novel relations between certain molecules had become inconceivable, and desires for creating fresh zones of proximity were cast aside. At the time, estrogen receptor research was of particular interest because these receptors had just been reported to behave in a manner that destabilized a dominant or "majoritarian" paradigm in neuroendocrinology. Estrogen receptors were traditionally thought to come in only one form. For a long time the field of endocrinology was not aware of the possibility that more than one type of estrogen receptor could exist. However, just before I started my research, the stable unitary identity of the estrogen receptor was displaced by the discovery of another nuclear receptor that bound to estrogen.[19]

This caused a minor endocrinological skirmish in its day, forcing estrogen receptors to be reclassified as either estrogen receptor-alpha (ERα, the "original" estrogen receptor) or estrogen receptor-beta (ERβ, the "other" estrogen receptor). I began my nomadic wanderings in search of estrogen receptors in the *in vitro* model of GnRH neurons, moving in and out of Petri dishes, in the shadows of transgenic politics, surrounded by scandals of scientific authority, exposed to the gender, race, and class dynamics of scientists in the laboratory, and faced with the anxiety of protein otherness on many different scales—basically a day in the life of a feminist scientist. All this was worth it, however, because I knew that by participating in the production of scientific knowledge on the body, I would have a chance to bring together feminist politics and marginalized or "minor" literatures in neuroendocrinology to form new lines of inquiry. Many scientists as well as feminist health activists have suspected for some time that estrogens present in hormone-based therapies and reproductive technologies may be doing more than simply managing the symptoms of menopause or regulating ovulation. I was excited to participate in the production of scientific knowledge that examined the possible direct neurological impacts of estrogen and could perhaps help to address some concerns around the design and use of estrogen-based hormonal contraceptives, hormone replacement therapies, and new reproductive and genetic technologies.

I recall the excitement that my colleagues and I shared when we first searched for estrogen receptors. We used molecular biology techniques such as subcloning and sequencing to find the genes for ERα and ERβ. We then carried out reverse transcriptase polymerase chain reaction (RT-PCR) experiments on *in vitro* GnRH neurons, in search of ER cDNA (complementary DNA). The idea was to show that the transcription and translation of these genes ultimately led to the expression of ER proteins in GnRH neurons. Not only did we find ERα cDNA coding for the gonadal hormone receptor that had been excluded from the mind's eye of neuroendocrinologists, but we also found ERα's other, ERβ.[20] We went on to conduct more molecular biology experiments such as Western blot analysis, to search for the expression of ERα and ERβ proteins. Both were found easily. Our results proved not only that GnRH neurons express estrogen receptors but that these receptors directly repress GnRH synthesis in these neurons. The life of these experiments set me onto a new line of flight and onto a series of interdisciplinary inquiries that continue to motivate me to this day. I faced and overcame scientific authority that had stacked the odds against this research and against the ability of estrogen and GnRH neurons to form more intimate zones of proximity. Although I did not have the vocabulary at the time to express myself in this way, I knew that I had witnessed a kind of ontological rupture that had moved my understanding of matter from one of fixity, stasis, and being to that of flexibility, change, and becoming. It was as if one night I had gone to bed with a materiality in which estrogen receptors could not be brought into closer proximity to GnRH neurons. The next morning, I woke up to an alternate materiality with new sites of play and unexplored biopossibilities for these biological matters.

Since I conducted these initial experiments, there has been even more reason to see that biologies are not fixed and to appreciate the value of moving toward playfulness and the power of the erotic in our search for biological knowledges. It has been further reported, for instance, that ERβ actually comes in not one but four different orientations or isoforms, including ER-β1 as well as ER-β2, ER-β4, and ER-β5.[21] A third type of estrogen receptor has also been isolated, known as GPER1 (also referred to as GPR30).[22] What is fascinating about this particular estrogen receptor is that it is a G protein-coupled transmembrane receptor (GPCR) and not a traditional nuclear steroid receptor. In GnRH neurons, which have been found to express this GPCR, this means that estrogen not only regulates nuclear transcription mechanisms, which happen on a timescale of hours, but that estrogen is also capable of eliciting rapid excitatory membrane-initiated actions, which can take effect within seconds to minutes. This membrane-bound estrogen receptor can directly modify GnRH neuronal activity and can trigger several different signal transduction pathways that are not affected by the traditional ERs that act as nuclear transcription factors.[23] Once again, this discovery of membrane estrogen receptors may be of particular interest to feminist health and reproductive justice advocates, as rapid and direct effects of estrogen can help to explain the long-observed "side effects" that many women experience while using estrogen-based contraceptives and therapies. The discovery of ERs in GnRH neurons, and the further discovery of G-protein coupled estrogen receptors in both GnRH neurons and other tissues throughout the body, serve as opportunities for feminists and molecular biologists to reach toward not just one but several shared objects of perplexity.

The life of this particular lab experiment continues to shape my efforts in theory-making. Beginning with my efforts as a scientist trying to become a feminist scientist, to a feminist scientist trying to become a women's studies scholar, this lab experiment has encouraged me to slip and grow in the midst of many disparate fields and has extended itself into an immensely productive experiment in interdisciplinary scholarship. It has provided me with the impulse to bring together scholarship that is typically separated by disciplinary distinctions such as the humanities and sciences, as well as by cultural distinctions such as concepts of the East and the West. *Molecular Feminisms* was written with several different audiences in mind. In addition to joining some ongoing conversations in the fields of feminist theory, postcolonial and decolonial studies, posthumanism, new materialisms, and science and technology studies (STS), this book hopes to provide a reflective space for both feminist scientists who wish to participate in bench research and the production of scientific knowledge as well as scientist feminists who are eager to use scientific research and data to inform their feminist analyses.[24] The past few decades of work in feminist STS has prepared us for noninnocent entry into lab spaces and participation in the production of scientific research. However, rather than following the more dominant women-in-science pipeline ideology, I argue that feminists might want to try to proceed in this noninnocent entry by exploring less common, marginalized, or minor modes of engagement with the sciences. My goal is to contribute to theory-making by creating conceptual frameworks that can be used to approach the lab bench, bring scientific research and data out of the lab, and revitalize how we think about bodies, biologies, and matter.

As a feminist scientist, I recall my interest and excitement after learning several feminist critiques of science. These critiques were eye-opening, compelling, and made great sense to me. Inspired by these critiques, I faced the challenges of actually trying to apply feminist epistemologies and methodologies at the level of practices at the lab bench. This was no easy task. While conducting research in the lab, the feminist scientist does not have a great deal of spare time, or the tools for that matter, to reflect upon and build connections between their love of science and their commitments to social justice. Facing this challenge has perhaps been the most productive and rewarding aspect of my intellectual career. Learning how to articulate this challenge into questions that make sense to both feminists and scientists has been an experiment in itself.

The scientist feminist who is interested in working with the sciences and scientific data may also benefit from learning more about the nitty-gritty practices of experimental biology. Recent projects have encouraged feminists to turn to questions of matter, and to animal, vegetal, and molecular bodies to develop more intimate treatments of materiality. Although new ontological gestures have revived feminist theory's engagements with the sciences and with biology in particular, without an effort to connect this theory to the theory and practices taking place at the level of the lab bench, these gestures run the risk of suffering a similar fate as that of poststructuralist feminist theory and earlier feminist critiques of science, which rightly or wrongly have been accused of working to restrict our access to the natural world. Worse still, what appears to be a growing tendency in some recent materialist scholarship to accept scientific knowledge at face value could in fact do feminism a disservice. As a feminist scientist, I offer some new perspectives and tools that one might use to approach biology from outside the lab.

I address some big questions that both feminist scientists and scientist feminists may have in common. How do we continue with science after the critiques of science? How do we work toward a biology that we desire? How are we to encounter matter? How can we bring questions of context with us when we do encounter this matter? How can we reconfigure the relationship between the scientific knower and what is to become the known? These and other such interrogations are visited several times, in multiple ways, throughout *Molecular Feminisms*. Admittedly, some of these interrogations are theory-heavy and draw from larger philosophical projects, but a genuine effort is made to make them relatable to the busy bench scientist. I articulate these interrogations by providing examples of the challenges one might face while trying to practice science as a feminist or practice feminism as a scientist. These examples provide tools that can be used for going into the sciences, developing methods, presenting our results and data, and facing ethical dilemmas in our research.

Part of the challenge of developing an interdisciplinary experiment involves starting with distinct disciplinary vocabularies, even when one is trying to address an idea or concept shared by these different knowledge systems. Building shared vocabularies takes a great deal of patience. Having said this, I encourage humanities scholars to work through the scientific experiments and data presented, and similarly I encourage scientists to push through the more philosophical aspects of the book. Chapter 1, "Biophilosophies of Becoming," is a particularly theory-heavy chapter. Establishing the book's philosophical framing, it provides tools for thinking about a broader conversation at the intersection of philosophy and science. Truly interdisciplinary work takes time, is full of failures, and can often leave everyone concerned unsatisfied. As excited as I am to bring this interdisciplinary conversation forward, I am fully aware that in my attempt to write a book for multiple audiences, I may have in fact written a book that is legible to no one, except perhaps myself—and even that is not guaranteed. What can I say? It's an experiment.

Feminism, Science, and the Politics of Knowledge Production

Feminists have a long history of intervening in the politics of knowledge production in ways that are specific to their time, location, and culture. They have argued that such factors as one's sex, gender, sexuality, race, ethnicity, class, age, abilities, location, and more influence who gets to conduct research, which questions generally get asked, and what knowledge is ultimately produced.[25] With today's fast pace of biotechnological developments, feminists are aware, more than ever, of the importance of bringing these critical perspectives into dialog with the sciences. There are many different types of feminisms, and although there is overlap, each has its own various forms of analytics. Among the different genealogies of feminism, the project at hand highlights some of the distinct approaches that exist between what has been referred to as liberal feminisms or feminisms of equality and what is often referred to as difference feminisms. Although these genealogies are no doubt messy and tangled, Elizabeth Grosz has provided an explanation of some of their distinctive features. "In place of the essentialist and naturalist containment of women," she explains, "feminists of equality affirm women's potential for equal intelligence, ability, and social value."[26]

Grosz suggests that for many, equality feminisms are generally motivated by a "logic of identification," which is "identification with the values, norms, goals, and methods devised and validated by men."[27] Alternatively, difference feminisms highlight women's differences from men. However, as Grosz further explains, "it is vital to ask how this difference is conceived, and, perhaps more importantly, who it is that defines this difference and for whom."[28] As she notes, it is important to understand that for these feminists "difference is not seen as difference from a pregiven norm, but as pure difference, difference in itself, difference with no identity."[29] These distinctions between feminisms of equality and feminisms of difference can be understood to roughly align with the philosophical and political approaches referred to respectively as "Molar" and "molecular" politics by Deleuzian scholars.

Following major feminist social and political interventions in the 1960s and 1970s, which in the US context were generally aligned along feminism-of-equality frameworks, many feminist scholars and activists turned their attention to the authority, validity, and impact of scientific claims. These claims were examined specifically for their role in supporting gender-based discrimination of women within academia, the home, and the workplace. For example, beginning in the late 1960s, feminists in the women's movement who were invested in participating in the production of scientific knowledge in the biological and reproductive sciences came together to form such groups as the Boston Women's Health Book Collective (1969), the feminist Self Help Clinic in Los Angeles (1971), and the National Black Women's Health Project (1984).[30] During this era feminists in the United States formed these organizations and developed their own knowledge bases in response to, and as an alternative to, decades of research on women's bodies, biology, and health that had originated from scientific disciplines and governmental institutions with long histories of excluding women and other minority groups as credible researchers and/or policy makers.

Beginning in the early 1980s, and working mostly in the US, Canadian, and European contexts, feminist philosophers of science, feminist historians of science, feminist sociologists and anthropologists of science, and several critical scholars contributed to the efforts of disciplinary and institutional change by developing highly sophisticated critiques of traditional and dominant forms of scientific research.[31] Many of these feminists made their interventions in the sciences by conducting in-depth critiques of the epistemological framings, methodological approaches, and language and metaphors commonly used in scientific teaching, research, and publications. These modes of critique are still relevant and operational today and continue to serve as important sources of feminist engagement with the sciences, particularly within the biological and life sciences. Some key aspects of these feminist critiques include (1) questioning the capability of achieving "pure" or aperspectival objectivity; (2) interrogating the impact of reductionist thinking; (3) problematizing the use of binary categories in the organization of observed biological and behavioral differences; (4) pointing out essentialist assumptions in scientific theories, specifically those that reinforce and promote biologically deterministic reasoning; and (5) questioning linear logic and oversimplistic models that move too easily from observations of correlation to explanations of causation. Many of these feminist critiques of science were also accompanied by practical suggestions for the diversification and democratization of science, both in terms of who should have the opportunity to participate in the production of scientific research and which ideas should be included in the pursuit of evidence-based scientific knowledge.[32] Put together, the work of early feminist health advocates and the strategies of critique developed by feminist academics trained in the humanities and social sciences have served as crucial cornerstones in the theoretical development and applied practices of the field of feminist STS.

In addition to humanities and social sciences–based scholarly engagements, in the early 1980s several notable feminists who were also scientists began contributing to the early formations of feminist STS. They made their contributions by first meeting the challenge of training and practicing in the "hard" sciences in academic and workplace climates that were more often than not hostile to women. Many of these feminists went into the sciences because of their love of biology, physics, or chemistry, and after having met the material challenges of working within these disciplines, they began sharing their experiences, critiques, and informed calls for change. Starting in the early 1980s, many feminist scientists shared their hands-on experiences of living and working within the sciences. They include, for instance, Margaret Benston, Evelyn Fox Keller, Ruth Bleier, Anne Fausto-Sterling, Lynda Birke, Sue V. Rosser, Ruth Hubbard, Donna Haraway, Lesley Rogers, Ursula Franklin, Bonnie Spanier, Karen Messing, Donna Mergler, and Karen Barad. Their critiques of the sciences, conducted from the "inside," were not meant to shut down feminist dialog with the sciences or discredit the sciences but were rather aimed at improving and furthering scientific knowledge in their respective fields.[33]

Feminist biologists working during this era were well positioned to critically analyze research and data, at the level of basic laboratory bench science as well as at the level of behavioral studies conducted on animal and human subjects in the clinical environment. They conducted intimate critiques of the multiple disciplines of biology while also working with animals, plants, microorganisms, and other biological materials in the lab. Their careful analyses were produced through firsthand experiences and insights of the specificities associated with scientific practices such as experimentation, statistical analysis of data, and scientific publishing. The skills these feminist biologists acquired while working within the sciences were hard-earned and gave them a degree of legitimacy that was required in order to critique and comment on the state of their particular discipline's understandings of bodies, biologies, and matter. They continue to serve as role models for generations of feminist scientists to come, because rather than shying away from the challenge of training and working within the hard sciences, or keeping their experiences in the sciences to themselves, these feminists shared their thoughts and made the inner workings of scientific research and knowledge production more transparent to others. Their efforts made it possible to know more about what one was getting into by signing up to become a feminist scientist.

Having said this, many of the early critiques of biological research and knowledge made by feminist scientists beginning in the 1980s were structured along liberal equality-based feminist frameworks. These critiques were in line with feminist critiques taking place at the time in humanities and social sciences–based disciplines such as philosophy, history, literature, political science, anthropology, and sociology. Rightly so, feminists during this era were primarily invested in questioning the dominant gendered paradigms operating in the traditional disciplines of the humanities and social sciences whereby women, or traits associated with femininity, were deemed as being inferior to men or masculine traits. Feminist scientists critiqued essentialist and deterministic modes of thinking and experimentation within the sciences, largely by addressing the epistemological and methodological biases apparent within their own particular areas of scientific expertise. However, these studies did not explicitly develop alternative epistemological or methodological approaches for conducting biological research that could be viewed as being nonessentializing or nondeterministic. In addition, not a great deal of attention was paid to the ontological assumptions and implications operating within their own disciplinary frameworks.

Despite this, it is misguided to suggest that feminists who took the trouble to train and spend time within the natural sciences, and particularly within the life sciences, did not develop their own intimate inquiries into the nature of matter or were not deeply aware of the importance of developing a *feminist ethics of matter*, even if that is not the vocabulary they used to describe their work. It is problematic to suggest that their critiques of scientific research were aimed at dismissing the contributions of bodies, biologies, and matter, or rejecting the knowledge that could be gained by the disciplinary fields of biology, genetics, and molecular biology. Interestingly, a similar accusation has been brought against feminist theory that has been influenced by poststructuralism and cultural theory, and unfortunately in some cases to all of feminism in general. By returning to the earlier work of feminist health advocates, women-of- color feminists, feminist philosophers, historians, sociologists and anthropologists of science, and particularly to the accounts of feminist scientists with the productive generosity of generational feminisms, we can begin to view this important work in new ways and thereby sharpen our current analyses in feminist STS.[34]

Much has transpired in both feminist political struggles and various areas of biological research since these early feminist engagements in the 1980s. The vibrant field of feminist STS has become an integral part of disciplinary training in women's studies, and for feminists who continue to train and practice in the sciences, feminist STS has come to serve as a toolbox, providing a compass from which one can navigate their own attempts at scientific knowledge production.[35] As the importance of feminist STS becomes more recognizable to scholars working in other fields, we are witnessing an explosion of interdisciplinary activity in an already interdisciplinary space. For example, feminist and queer scholars working at the intersections of philosophy, poststructuralist theory, cultural studies, literary studies, and psychoanalysis have also turned their attention to STS. They have brought with them the skills of questioning dominant metaphysical traditions and are imagining new ontological orientations and ethical gestures that can be used to think more critically about our relations to matter and the world around us.[36] They offer the experience of developing alternative frameworks for thinking about questions of difference, identity, and representation.

Many of these ontological and ethical reorientations and gestures may not be entirely new or unfamiliar to feminist scientists or feminist STS scholars. For instance, the question of the relationship between the scientist and their "object" of study has been at the heart of several feminist reflections on science and scientific method.[37] What is new is the interdisciplinary and shared vocabulary that is developing around common questions related to matter, ethics, and knowledge-making practices thanks to the commingling of theories and vocabularies between various fields. Over the past decade these intellectual collaborations have led to an exciting burst of scholarship found in feminist STS. This long trajectory of feminist materialisms—starting with feminist health and reproductive justice activism, to early feminist critiques of science, to current-day interests in feminist theory regarding the ontological status of matter—has brought me to write this interdisciplinary book about feminism, molecular biology, and the importance of theory-making both inside and outside of the lab.

Why Molecular Feminisms?

Other than my own interest in molecular biology research, and the obvious word play between "molecular biology" and "molecular politics," what claim or distinction am I trying to make by turning to the molecular? Although I am in no way interested in dismissing current feminist STS projects that are aligned with women in science and feminisms-of-equality projects, I am invested in theory-making that can emerge from using philosophical and political approaches that turn to more marginalized or underplayed ideas, literatures, and thinkers in both feminism and the sciences. What can happen at the intersections of feminism and science when we look to less familiar figures, both human and nonhuman, for our theory-making? I must admit that it is my training as a scientist (specifically as a molecular biologist), and not my interest in feminist theory, that first brought me to think about the difference between molar and molecular approaches. It is the years of making chemical solutions in the lab, learning about the behavior of molecules, and studying the microdynamics of signal transduction pathways that have led me to gravitate toward the molecular.

In their collaborative text *A Thousand Plateaus: Capitalism and Schizophrenia*, Deleuze and Guattari have described those modes of thinking and politics that draw upon philosophies of being, stasis, and identity as being "majoritarian" or "molar" in their approach. Alternatively, they describe those tactics that build upon the ideas of becoming, change, process, and events as being "minoritarian" or "molecular" in their approach. Deleuze and Guattari repeatedly emphasize that these tactics are not opposed to each other, but rather that they can be distinguished by their orientations to matters of scale.[38] In *A Thousand Plateaus*, Deleuze and Guattari attempt to move away from a Platonic metaphysics. To do so, however, they know it is necessary to account for the presence of forms and substances, which they attempt to do by suggesting that forms and substances are "generated by intensive processes rather than imposed on intensive processes from without."[39] During this treatment of forms and substances, Deleuze and Guattari make a distinction between the molar and molecular. They state:

> It is clear that the distinction between the two articulations is not between substances and forms. Substances are nothing other than formed matters. Forms imply a code, modes of coding and decoding. Substances as formed matters refer to territorialities and degrees of territorialization and deterritorialization. But each articulation has a code and a territoriality; therefore each possesses both form and substance. For now, all we can say is that each articulation has a corresponding type of segmentarity or multiplicity: one type is supple, more molecular, and merely ordered; the other is more rigid, molar, and organized. Although the first articulation is not lacking in systematic interactions, it is in the second articulation in particular that phenomena constituting an overcoding are produced, phenomena of centering, unification, totalization, integration, hierarchization, and finalization.[40]

Inspired by their interests in the natural and physical sciences, Deleuze and Guattari draw distinctions between molar and molecular thinking by turning to and drawing parallels with geology, chemistry, and biology.

They have described majoritarian politics as having molar tendencies, because these approaches often deal in identity-based, territorialized, organized, originary, and often privileged terms.

The term "molar" in chemistry refers to a unit of concentration (known as molarity) that is equal to the number of moles of a substance per liter of a solution. A mole, in turn, is a chemical mass unit of a fixed number (6.022×10^{23}) of molecules or atoms of a substance, also known as Avogadro's number. This representation of a group of molecules or atoms that come together to form one entity is what Deleuze and Guattari allude to in their use of the term "molar." Alternatively, they describe minoritarian approaches as molecular tendencies, not because they belong to a minority group or that they operate only at a subcellular level but because they entail those ethical actions and ontological maneuvers that work to deterritorialize our thoughts. As Eugene Holland has explained:

> There are several ways of approaching the relations between molar and molecular. One is in connection with the articulation of content and expression. As we have seen, a substance can take liquid form on the molecular level, and then get transformed into a crystal on the molar level: water vapor becomes a snowflake. Notice that molecular and molar are relative terms: when individual snowflakes combine to form a snowdrift, or a snowman, it is now the snowflakes that constitute the molecular level, while the snowdrift and snowman are molar. . . . The recourse to statistical probabilities may be what gives rise to the false impression that the difference between molar and molecular is a matter of size, when in fact it is more a matter of perspective.[41]

My reason for turning to the molecular, and to questions of *becomings*, is directly related to becoming a feminist scientist, working at the intersections of reproductive neuroendocrinology and molecular biology, and the quandaries regarding matters of perspective that these experiences have produced.

This distinction between major/minor and molar/molecular politics and matters of perspective is also found in the work of new materialists but is best expounded by Elizabeth Grosz and Rosi Braidotti. Both Grosz and Braidotti have reflected carefully on Deleuze and Guattari's philosophical interests and are aware of the valid feminist criticisms of molecular concepts such as *becoming-woman*.[42] However, they have also created a space for lively exchange between the work of Deleuzian ethics, feminist theory, and feminist STS. Their contributions to feminist theory and feminist STS—particularly their ontological and ethical reflections on questions of difference, sexual difference, and molecular politics—have served as crucial points of reflection for me. Grosz, for example, has argued that the molecular is a way of thinking through difference in terms of difference in and of itself. "If molar unities, like the divisions of classes, races and sexes," she writes, "attempt to form and stabilize an identity, a fixity, a system that functions homeostatically, sealing in its energies and intensities, molecular becomings traverse, create a path, destabilize, enable energy seepage within and through these molar unities."[43] Interestingly, Grosz cautions that molecular projects such as those aligned with difference feminisms might often appear to reify differences and work against liberal feminisms committed to equal rights.[44] Despite this fact, she argues that molecular projects are necessary in order to think about difference, particularly to think about sexual difference through multiplicities rather than what Luce Irigaray has identified as a logic of the Same.[45]

Explaining the distinction between molar and molecular projects, Braidotti suggests that "the 'Molar' line" is "that of Being, identity, fixity and potestas—and the 'molecular' line—that of becoming, nomadic subjectivity and potential."[46] We can look, for instance, to the impact of identity-based molar politics, which in the case for humans has led to many advancements made by women's rights, civil rights, gay liberation, and disability rights movements. The ability to claim membership within a group that is marked as a stable identity—or as Gayatri Chakravorty Spivak has explained, to be able to strategically invoke an essentialized identity such as calling oneself woman, lesbian, transgender, intersex, or disabled—can carry much political import.[47] It is therefore crucial to make clear that by turning to ideas of becoming, movement, change, and intensities—what I refer to as *molecular feminisms*—I am not attempting in any way to discredit or devalue majoritarian naming practices or molar identity-based representational politics. Also, I am not suggesting that one must necessarily have to choose one mode of thinking about and approaching the world over the other. Rather, I want to acknowledge, as Deleuze and Guattari point out repeatedly in *A Thousand Plateaus*, that being and becoming coexist and even work to coproduce each other.[48]

Similarly, we could keep in mind the fact that stoloniferous plants grow by both extending stolonic shoots and establishing adventitious roots at its nodes. Yet, I also want to acknowledge that our habits of logic have limited how it is that we most often pursue knowledge about ourselves and the world around us. We have been limited by an all too familiar mode of questioning and reasoning through molar modes of being. This has been the case in the vast majority of encounters between feminism and biology.

Approaching Matter through New Lines of Flight

I have always had a passion for the natural sciences. *Molecular Feminisms* is written from the perspective of a feminist STS scholar who had the benefit of learning from feminist critiques of science; in fact, because of (and not in spite of) these critiques, I went into the "hard" sciences purposefully at a time immediately following the "science wars."[49] Encouraged by feminist activists and scholars around me, who themselves were deeply involved in identity-based women's rights and reproductive justice movements, I pursued my doctoral training in molecular and reproductive biology in order to gain expertise in the scientific theories and practices that were directly related to our knowledge of women's reproductive health. I was exposed to some crucial scholarship in feminist STS while I was training in the lab, including theoretical interventions mapped out by, to name a few, Donna Haraway's situated knowledges and cyborg manifesto, Helen Longino's outlines for socially just science, Sandra Harding's concept of strong objectivity, Emily Martin's suggestions for new ways of conducting reproductive biology research, and Banu Subramaniam's metanarrative on science and scientific method.[50] Bonnie Spanier's *Im/Partial Science: Gender Ideology in Molecular Biology* was particularly eye-opening for me, as the work provided a methodology for revealing gendered paradigms in molecular biology research.[51] These were the feminist theoretical interventions and methodological tools I took with me to the lab bench while training to become a scientist.

However, I wasn't prepared for the journey of ontological and ethical reorientations that I was to embark upon as a direct result of my training in molecular biology. Nor could I have anticipated the reactions I would encounter from my feminist colleagues in women's studies departments, women's health movements, and reproductive justice organizations as a result of these reorientations. I went into the biological sciences in the first place because of the strong impulse *not to ignore questions related to matter* that I had learned from feminist philosophers of science, women of color feminisms, postcolonial studies, and women's health activism. I knew very well that I was making a noninnocent entry into the lab. Once I got there, I realized that I was somewhat on my own in my attempts to bring my feminist interests in matter and materiality, which were directly related to women's reproductive health, together with the everyday, nitty-gritty practices of molecular biology. As a feminist scientist, I wanted to participate in the production of scientific knowledge, but as a result, I found myself asking a series of challenging and difficult questions that I had to face head-on if I wished to learn more about matter through the practices of molecular biology. These questions were multiple and varied, and when I tried to bring the challenges of doing bench science back into conversation with my feminist peers who worked outside of the sciences, my questions were often met with a fair bit of confusion, if not alarm. The reactions I received after posing such questions as "Should feminists clone?" (a genuine question I needed to ask as a molecular biologist) indicated that my ontological and ethical reorientations had led me to venture out a little too far.[52]

Learning the everyday practices of bench science in molecular biology taught me a number of invaluable lessons. Before I was introduced to the work of Isabelle Stengers, I had a hunch that in order to create meeting places for meaningful interdisciplinary conversations and opportunities for imagining joint biological and technological futures, both molecular biologists and feminists could benefit by learning about each other's practices.[53] While training to become a molecular biologist, I was genuinely interested in learning how to work with biological matters through the practices specific to this field of scientific research. During this process, did I take with me the feminist critiques of science that taught me how to recognize sexism in the sciences? Yes. Did I learn to recognize and name institutional racism and how it operated in my university and department? Yes. Did I register the gendered, racist, classist, and ableist language and paradigms that surrounded me in molecular biology or reproductive physiology textbooks, lectures, or scientific articles to which I was exposed? Absolutely, yes! Did I see a relation between those who wore a white lab coat, their pursuit of pure objectivity, and the drive within the sciences to erase matters of context? Repeatedly, yes. Despite these active recognitions, however, I did not approach the actual nitty-gritty everyday practices of molecular biology and reproductive neuroendocrinology laboratory techniques and experiments through a sense of irony or distrust. Nor did I intentionally plan to become a scientist and spend years of my life learning the incredibly complex practices of my field only to simply dismiss them or to "bring science down" as it were. Constantly aware of the importance of thinking with molecular biology, I also registered the constraints and difficulties posed by this research. Ultimately, I was driven by the desire to proceed.

As a direct result of my participation in molecular biology research, I learned how to ask informed questions about the scientific method, objectivity, reductionism, and how to develop informed feminist critiques of science. I also began to articulate a new line of philosophical inquiry for myself into the nature of knowledge, the nature of biological matters, and the nature of being. I have always been drawn to the minor literatures of molecular biology. Years later, after having a chance to more carefully read the work of other feminists who also trained in the sciences (such as Haraway, Stengers, Barad, and Subramaniam), I am able to express my reasons for entering into the sciences in the first place. I feel more capable of expressing my earlier insights into the importance of learning from disparate practices, in terms that are more familiar or recognizable to my colleagues in the humanities and social sciences. I am better able to express the interdisciplinary impulse that I have had to bring feminism and molecular biology into conversation in less commonly explored ontological, epistemological, methodological, and ethical terms. I am now able to truly appreciate the value of learning *how to encounter* the varied materials and practices of a discipline that is not one's own, through the productive lens of "shared perplexities."[54]

Most important, my original hunch about the importance of learning to appreciate and respect the practices of disparate disciplines has led me to reframe how we might think about molecular biology and feminism together, how we may be able to work together to create new lines of flight to think about the world that we inhabit, and how we can produce knowledge about that world. This is why I am drawn to molecular projects, both in molecular biology research and feminist politics, that can present alternatives to dominant modes of thinking.

A Note about Methodology

Grasses, apparently, are notorious for crossing taxonomical boundaries but are generally recognized to grow in three ways: as cespitose grasses (grass that grows in bunches with straight roots), turf or sod grasses (grasses that grow by spreading their roots outward horizontally), and matgrasses (which fall somewhere between the other two grass formations). The roots of cespitose grass grow in clonal patterns, whereas all other grass species grow and extend their roots either as rhizomes or stolons. Like the rhizome, the stolon is not just a taxonomical classification but also a strategy for plant propagation. Both stolons and rhizomes form internodes from where new root systems can begin. Stolons, which are referred to as "runners" for their ability to move horizontally above ground, have the additional capacity of serving as "foraging organs for light."[55]

Stolons have the ability to extend runners in multiple directions and also the capacity to sense their surroundings. As an interdisciplinary project, *Molecular Feminisms* goes by the way of the stolon and the stoloniferous plant, creating runners along different directions, foraging in and out of properly defined disciplinary formations—all the while aiming to connect and contribute to the field of feminist STS. If extending toward and trying to touch shared objects of inquiry counts as a methodological strategy, this book does just that by attempting to extend runners between feminist theory, postcolonial and decolonial theory, posthumanist ethics, new materialisms, philosophy of science, and molecular biology. While there is a clear emphasis on the foraging strategies used by the stolon, the fact that stoloniferous plants also grow nodes with roots that go down into the ground is not ignored in this work.

To sketch the methodological approach that ties this book together, I want to briefly recall an exchange I had with a well-known sociologist and STS scholar soon after completing my PhD and starting my tenure-track position. During that encounter this very generous colleague cared to ask me about my work and was curious about how and why, with a PhD in molecular biology and neuroscience, I had ended up in a women's studies department. After briefly listening to me fumble my way around saying that I wanted to bring feminism and science together to generate new kinds of conversations, he quickly summed up my methodological allegiances by saying, "Oh—so you're an ethnographer! You study scientists in the lab!" If being an ethnographer means observing and interpreting the actions of the people and culture around oneself, then yes, I am an ethnographer. But to be clear, this book in no way meets any standards of a proper ethnography. While I was training to become a scientist, and conducting my own experiments in a lab, I was not systematically taking notes on my colleagues around me or documenting how as scientists we come up with our hypotheses and conduct our scientific experiments.

Rather, I went deep into studying the practices of molecular biology, and through a slow and sometimes imperceptible process, perhaps much like watching grass grow, I became an expert in learning how to spot both the challenges and possible points of entry for creating interdisciplinary work and shared moments of perplexity. Having been trained in the biological sciences and having now spent a significant amount of time as a feminist STS scholar, I have come to the conclusion that the hardest task of interdisciplinary scholarship entails not only learning how to come to the table but also knowing how to assemble a table that will actually support joint conversations. This process takes time, and the results are not always immediate. A methodology of reaching out and making connections requires the slow and painstaking work of developing shared vocabularies and respect for distinct and sometimes quite disparate practices. It takes time to learn how to frame one's questions in a way that they can actually be heard from another disciplinary standpoint.

Another important point on the methodological framing: as mentioned previously, I write as a feminist scholar who conducted graduate training in the sciences after the science wars. As a result, this book has been deeply informed by debates between social constructivism and positivism and the intellectual fallout that transpired between poststructuralists and scientific realists. In the framing of *Molecular Feminisms*, science and feminism are treated as being co-constituted. This interdisciplinary text aims to bring together the work and methodologies of diverse groups of scholars to think about overlapping sets of questions regarding the nature of scientific inquiry, the nature of the relation between a knower and the known, and the nature of feminist political movements. Whether articulated in a similar way or not, both groups of stakeholders—molecular biologists who may not necessarily identify as being feminist and feminists who are not trained scientists—know that it matters how one orients oneself toward an object of inquiry and that the practices of scientific observation carry with them the weight of great political impact.

By taking into account our orientations toward a blade of grass, it could be said that I am also trying to work "from below"—or in the case of using stolonic strategies, from along the surface of a flat plane or ground. Feminist philosopher Sandra Harding has taught us the importance of using feminist, postcolonial, and decolonial theories and practices to approach the "sciences from below."[56] What I am in search of, however, are ways to think about feminism and molecular biology through minor literatures and modes of molecular thinking, as Harding reminds us, by "keeping both eyes open."[57] In chapter 1, I develop biophilosophies of becoming, which is about learning how to start thinking not just from below but from "far far" below the usual human point of entry and view. It is about learning how to take ourselves down a notch or two, how to come to eye level with the stolon, and how to become like the blade of grass in order to make a more communicating world. The hope is to draw from disparate theories and practices that can become part of our everyday politics. These practices exist both inside and outside of the lab and can help us, as Claire Colebrook has stated, to "grasp all the inhuman perceptions and forces beyond the order of our point of view" that contribute to flourishing.[58]

I am also aware of Harding's methodological prompting of approaching the sciences from below in order to locate the Eurocentrism in our own work.[59] It is indeed the case that as a scientist trained in the global North, I have been exposed to those canons deeply embedded within male-dominated and Eurocentric knowledge traditions. In addition, my limited language skills block me from reading and writing about both science and feminism in any other language than English. I acknowledge these factors up front as methodological limitations of the work presented here. To address this point, I attempt to reach out toward a number of different thinkers, a great many of them from South Asia, to reflect on diverse accounts of the world around us and diverse accounts of how we can come to know this world. These accounts do not aim to displace modern scientific understandings or theories of origin and evolution of the universe. Rather, I draw from scientists and philosophers whose work has been marginalized within both the sciences and feminist theory to think more broadly about questions of ontology, epistemology, and ethics and how these may add to our theories of the universe and its inner workings. Just as important as it is to recognize that I am a scientist trained in the global North, as a diasporic South Asian, born and raised in Canada, with Bengali immigrant parents who grew up during India's partition, it is important to recognize that there are also certain anticolonial and postcolonial logics that have been woven into my methodological approach. Much like the character and scientist Sneha in Subramaniam's essay "Snow Brown and the Seven Detergents," I know very well that ignoring, whitewashing, or cleansing these logics from my intellectual palette is impossible, not to mention undesirable.[60]

Many of the figures I draw from are purposely Bengali (including Baba Premananda Bharati, Rabindranath Tagore, and Jagadish Chandra Bose), presented as anticolonial figures, who as colonial subjects under British rule in India were actively trying to communicate alternative ontologies and ethical approaches to matter and the world around us through religion, literature, and science at the turn of the twentieth century.[61] As Ashis Nandy wrote in his biographies of the scientist J. C. Bose and the mathematician Srinivasa Ramanujan, both British colonial subjects in India at the turn of the twentieth century, "How much has science lost by its mechanistic and physicalistic concepts of the universe, how much by its denial of all alternatives to the scientific culture of the industrialized world? How much has the Newtonian idea of a world machine contributed to the ethical predicament of modern science, to its role in fostering human violence, and in violence towards the non-human environment?"[62]

My intention behind creating runners between these figures is not to develop a project in comparative science or comparative philosophy. Nor is it to create a divide between the "East" and "West" and draw from Eastern traditions in order to dismiss Western traditions (or vice versa), as if these traditions and cultures could even be divided and determined so simply. I certainly do not provide in-depth biographical accounts of these thinkers or spend a sufficient amount of time reflecting on their full philosophical contributions and anticolonial positions. That would be a different book. I turn to these anticolonial figures for the minor literatures they developed in their own ways, and use these literatures to reframe majoritarian thought found in both science and feminism. Lastly, this book does not intend to use molecular biology research and methods to dismiss feminist concerns or, alternatively, to use feminist theories and methodologies to poke holes into molecular biology research. Rather, at the core of this inquiry is a deep curiosity that wants to see what is possible when two modes of thinking, simultaneously distinct yet in multiple ways co-constituted, are encouraged to interact reciprocally. What becomes possible when two modes of inquiry are allowed to move closer toward each other through unexplored processes of desire and produce new ways of thinking? *Molecular Feminisms*, like the stolon, is a strategy for *becoming-minoritarian*, *becoming-molecular*, and *becoming-imperceptible*.

Overview of Chapters

Chapter 1, "Biophilosophies of Becoming," establishes the theoretical framing of the book. It defines and deals with key questions of ontology, epistemology, and ethics that drive feminist, postcolonial, and decolonial STS interrogations of scientific knowledge production. The chapter asks us to reflect on how it is that we orient ourselves toward a diverse range of matters, starting with a blade of grass. It questions the nature of nature and what status we as humans have given to others, humans as well as nonhumans. It addresses a metaphysics and humanism that has influenced both our philosophical and scientific approaches to encountering the world around us and argues that in addition to spending our creative energies toward identification and classificatory practices (the values of which are not dismissed), we must also ask what matter, molecules, bodies, and organisms such as grass *are capable of doing*. This line of questioning opens us to the opportunity of developing *biophilosophies of becoming* that rethink matter in terms of flux, motion, and capabilities, and reframe biology in terms of events and processes. New ontological, epistemological, and ethical reorientations must also be contextualized and therefore I turn to postcolonial and decolonial STS approaches to consider the broader implications of knowledge-making practices. Building upon connections that can be made between feminist, postcolonial, and decolonial approaches, biophilosophies of becoming are more thoroughly articulated in terms of the qualities of (1) changefulness and nonhuman becomings, (2) kinship and hylozoism, and (3) univocity and immanence.

Chapter 2, "Microphysiologies of Desire," visits a long tradition of intimate ontological, epistemological, and ethical considerations made by feminist scientists and feminist philosophers of science regarding the relationship between the knower and the known. It acknowledges generational feminist materialisms and revisits questions that are familiar to feminist scholars—namely, how we can continue with science after the critiques of science and how we can reconfigure the relationship between the scientific knower and the known. The chapter explores earlier interventions made by Barad, Patti Lather, Stengers, and Haraway. By bringing together the scientific work of Bose and recent scholarship in posthumanist ethics, I examine what the implications would be of shifting our ethical stance from a transcendent understanding of "responsibility" toward the other, to a more immanent awareness of the "ability to respond" to the other. I offer *microphysiologies of desire* as strategies that bring us closer to a biology that we desire. Inspired by Evelyn Fox Keller's work on the life and scientific research of Barbara McClintock, I playfully reconfigure McClintock's famous dictum "a feeling for the organism" to "feeling *around* for the organism" to lay out a more immanent rather than transcendent orientation that can be put into practice in the lab. The aim is to speak to those feminist scientists and those scientist feminists who, in the wake of recent material turns, are eagerly looking for ways to participate in the biological sciences and work with and beside molecular actants in the lab.

Chapter 3, "Bacterial Lives: Sex, Gender, and the Lust for Writing," outlines key arguments in relation to poststructuralism, feminist theory, and posthumanist ethics. Following the lives of bacteria and the creativity and labor of bacterial writing, the chapter explores what new politics become possible when we think with members of a domain that have been noted for their qualities of changefulness and nonhuman becomings. Thinking alongside biophilosophies of becoming in relation to bacteria, the chapter confronts three binaries that are commonly used by scientists and feminists—namely, sex/gender, biology/culture, and matter/language. Beginning with the rich body of work left behind by the biologist Lynn Margulis, I trace how our understanding of bacterial sex has created new feminist theories and treatments of sex. I take note of the confusion that often exists between the terms "sex" and "gender" in biological research. Using the materiality of bacterial writing, reading, and modeling, the chapter illustrates the importance of including analytical frameworks in our theory-making that have been informed by decades of scholarship in feminist, postcolonial, and decolonial STS. Emerging from political movements that are invested in social justice, I argue that these frameworks not only allow us to posit new ontologies but also help us with the difficult task of putting these ontologies into practice.

Chapter 4, "Should Feminists Clone? And If So, How?," proposes a model for feminist inquiry from *within* the sciences. This model has the potential for broad use in the natural sciences but is most directly applicable to feminist scientists working in molecular biology labs. Michelle Murphy has suggested that women's health activists in the 1970s showed the "most sustained efforts to practice science as feminism."[63] Readers familiar with my earlier work will know that I also view my entry into and contributions to the sciences as expressions of my feminism. For this reason, and for the greater part of my feminist STS career, I have been committed to developing practices that are relevant to the feminist scientist in the natural sciences. After feminists learn the very important critiques of science, or learn to make astute ontological and ethical claims, they still need to directly engage with the sciences and the practices of scientific knowledge production. This can only take place when we spend time developing practice-oriented approaches in feminist STS and methodological interventions for the feminist scientist working in the lab.

Using the technology of subcloning, my goal in chapter 4 is to extend the scope of how we think about practicing science as feminism or feminism as science today. Theoretical interventions can emerge from both inside and outside of the lab, often as a direct result of the everyday nitty-gritty practices of both feminism and science. Formulating new feminist or molecular politics from a scientific protocol, a biological actant, or even a machine used in an experiment can be difficult. However, when we frame these efforts through biophilosophies of becoming, and through the qualities of hylozoism and kinship in particular, new modes of knowledge production become possible. The model of scientific inquiry proposed is based on a "mutated subcloning protocol," designed to acknowledge that a feminist scientist is accountable to multiple communities of knowers, which can lead to several tensions and dilemmas in their research. The model allows the researcher to take these dilemmas into account and move forward with a research agenda by formulating their questions in the lab in contexts that are most relevant to their own social justice commitments. The chapter details my own experience of conducting molecular biology research in a wet lab. Delving into my scientific research and discoveries in molecular biology and reproductive neuroendocrinology, I reflect on the tensions of using recombinant DNA technologies to search for estrogen receptors in GnRH neurons and the experiments that brought me to some of my current ethical and ontological queries.

Chapter 5, "*In Vitro* Incubations," focuses on new developments in the field of synthetic biology and explores how synthetic lives such as minimal genome organisms force us to reconsider the boundaries of interdisciplinary thinking. Starting with my experience of working in a molecular biology lab, I highlight several questions that I only learned how to ask by working with and paying careful attention to an *in vitro* neuronal cell line. Two main challenges face feminist STS scholars today—namely, addressing the question of what constitutes life and the living in this age of synthetic biology, and figuring out how to respond to and deepen our human entanglements with these lives by paying attention to questions of context and social justice in this molecular era. Turning to the qualities of univocity and immanence, the chapter explores how biophilosophies of becoming can present a different set of tactics that can be used for reaching out toward and responding to synthetic lives.

Following Deleuze and Guattari's idea of the machinic assemblage, both molar and molecular approaches are employed to understand the monstrous couplings that bring together digital DNA, humans, DNA synthesizers, Petri dishes, bacteria, yeast, and more. Synthetic biology has created a machinic assemblage that moves through three phases or strata of a new kind of synthetic life cycle, including the inorganic, organic, and social. In the inorganic strata we see the important role that the central dogma in molecular biology has played as a far-reaching paradigm, giving material form to minimal genomes. Through the organic strata we see the beginnings of expressive life in minimal genome organisms whereby they learn to communicate with each other and their external world. The last stage of the life cycle is examined through the social or alloplastic stratum, where we see how our machinic assemblage modifies the external world. Using the tools of postcolonial and decolonial STS, and through a series of vignettes, I analyze how synthetic lives simultaneously are produced by and work to produce social institutions and behaviors that are undoubtedly human but are not limited to the human.

1

Biophilosophies of Becoming

> The molecule's journey backward to Love is made by a very circuitous path. That path leads through the process of opening one by one the passages of its composing principles. By passages is meant channels of communication and sympathy with the main laws and vibrations of the working of the universe. The first step the molecule takes, in this return-journey, is by opening the passage of one principle, the sense of feeling, and becoming a blade of grass.
>
> —BABA PREMANANDA BHARATI

> Every time I walk on grass I feel sorry because I know the grass is screaming at me. . . . Basically everything is one. There is no way in which you draw a line between things.
>
> —BARBARA MCCLINTOCK

> In the world's audience hall, the simple blade of grass sits on the same carpet with the sunbeam and the stars of midnight. Thus my songs share their seats in the heart of the world with the music of the clouds and forests.
>
> —RABINDRANATH TAGORE

Becoming a blade of grass is perhaps an odd place to begin the task of developing an interdisciplinary project. Add to that the vision of screaming grass and the idea of a blade of grass sitting on a carpet in a hall as an audience member, and now there may be sound reason to flip back to the cover of this book and reread its title. Indeed, it is *Molecular Feminisms: Biology, Becomings, and Life in the Lab.* Starting this chapter by thinking about blades of grass that grow under our feet, or grass that grows on carefully manicured lawns if you happen to live in a concrete jungle, might appear at first to have very little to do with either molecular biology or feminism, let alone theories and practices in and out of the lab. However, the space where these matters all meet has everything to do with how we think about, and approach, the blade of grass.

Those of us who have had intimate access to grass—whether in our backyards, public parks, savannahs, or in actual grasslands—may be able to recall the smell of freshly cut grass or the feeling of cool blades under our feet. Perhaps you can remember sitting on a soft bed of grass while watching a faraway sunset. Some of us may even recall pulling out handfuls of grass only to find a shallow mess of tangled roots growing together. Alternatively, the thought of grass may conjure for you less idyllic reflections and instead raise concerns regarding the threat of deforestation, the development of genetically modified grass to provide even green turfs for playing golf, images of overly trimmed and herbicide-treated lawns that have been produced at high environmental costs simply for the benefit of human notions of "natural" beauty, or perhaps even remind one of old colonial English gardens and ornamental grasses that were planted and pruned outside of England in order to "civilize" supposedly backward and barbaric lands. Each and every one of these reflections is valid.

What the chapter epigraphs would have us consider, however, is that perhaps our thoughts on grass—whether we are pleased by its presence or not—may not just be about contemplating its role as a silent backdrop to our outdoor excursions, or considering the historical and political factors that go behind our enjoyment of picturesque green vistas. Rather, the epigraphs suggest that the thought of *becoming a blade of grass* or getting closer to grass is more about acknowledging the desires, voice, and status of grass. They raise the question of how we as humans act toward and respond to something as simple as grass, and ultimately how we might orient ourselves to a more diverse range of matters that constitute our own bodies, lives, and the world around us.

While taking pause to consider our relationship with grass, we may be inclined to start by ruminating on our own human status compared to that of grass and our grand place among all of nature's organisms and elements. The organisms and elements of nature referred to here include nonhuman animals, plant life, microorganisms, molecules, and even those rocks, minerals, and inorganic matters all around us that have been deemed inanimate or nonliving. By beginning with a blade of grass—or better yet, with molecules that love and that work toward becoming a blade of grass, with grass that has the capability to scream, or with a blade of grass that holds equal footing with the sunbeams and stars—we may be prompted to go even further and pose some difficult questions regarding not only the nature of our human encounters with the environment but also questions regarding the nature of nature itself.

We might, for instance, start to consider more seriously the possibility that the world is perceived and actualized not only by humans but also by nonhuman beings, organic and inorganic, animate and inanimate, living and nonliving. We might begin to pay more attention to and give more credit to the intricate and perhaps even imperceptible forces that push and pull, or desires that are enacted or expressed by nonhuman organisms, molecules, and elements, which undoubtedly contribute to the building up and breaking down of our universe. We might begin to consider more carefully the assemblages and entangled subjectivities that are bound to form between those who claim to observe this nature (never innocently) and those objects in nature, which are being observed (never passively). We may further start to question the limitations of what has come to constitute the human subject as an "autonomous agent" in a liberal humanist sense in our societies, who through a series of moral and legal codes has been given the right to either observe or the choice to be observed. We may start to wonder how our need for defining such a unified subject has undoubtedly resulted in the articulation of those who have come to constitute that subject's "other," both human and otherwise.

Where these questions regarding the nature and status of humans, human others, nonhuman others, subjects, objects, knowers, and the to-be-known lead us, is precisely where the disciplines of philosophy and science have always met and where they have also promptly become entangled. Some may argue that the boundaries between philosophy and science have never been clear. It is precisely at those spaces of unclear boundaries and entanglements between philosophy and science where feminist, postcolonial, and decolonial scholars have dedicated a great deal of their creativity over the past few decades through their analytical, theoretical, organizational, and material contributions. They have long known the high stakes involved in the not-so-simple acts of observing nature and producing knowledge. Many have pointed out that not all knowledge systems are considered equal and that in some dominant traditions of philosophy and science, not everyone is considered capable of being a knower, even when the knowledge that is produced relates directly to their own lives and experiences.

Feminist, postcolonial, and decolonial scholars have pointed out the cost of coming to know the world if this knowledge is sought and obtained only, or primarily through, the limited scope of "modern" or what is often referred to as "Western" traditions of philosophical and scientific inquiry.[1] These traditions of philosophy and science, which are generally traced back to the era of the classical Greek philosophers Plato (428–348 BC) and Aristotle (384–322 BC) and the European Renaissance (the fourteenth through the seventeenth centuries), are defined by a metaphysics that made very clear distinctions between the inferior ontological status of raw matter compared to the superiority of that entity which had the capability of moving from raw matter to assuming a form. Plato treated the body as a mere vessel for the soul.[2] This distinction operates in Cartesian dualisms that further separated "man" from nature and the mind from the body.[3] This transcendent distinction and relationship between form and matter, or the soul and the body, made its way into the Enlightenment (the seventeenth through the eighteenth centuries). Ultimately, this metaphysical framework of divisions, distinctions, and discontinuity has led us to the many philosophical and scientific approaches that we are taught and that we observe today, particularly in public and private institutions of higher learning. In addition to seeing the world through the lenses of discontinuity, transcendence, and dualisms, these approaches have been further limited, as some philosophers and historians of science have argued, by their epistemic reliance on reductionism and claims of achieving aperspectival objectivity.[4]

Feminist, postcolonial, and decolonial scholars have argued that these traditions in philosophy and science can be further characterized by the liberal humanist principles that reside at their core.[5] They have suggested that this form of humanism has led to the creation of hierarchies of being, where greater or lesser values are placed on different lives (including human) and expressions of being. Such dominant philosophical and scientific traditions are built upon the problematic belief that there is a singular and easily discernable positive teleology of thought and progress that is somehow unique to the West, and that this entity, "the West," is derived from a history and materiality that is distinct from and superior to "non-Western" thought and cultures.[6] These scholars are critical of the colonial and imperial-inflected definitions of what has come to constitute modern philosophical and scientific thought versus the so-called uncivilized or "premodern" traditions of knowledge production.[7]

Together, the disciplines of both philosophy and science, as perceived in "the West," have come to form a legacy whose modes of inquiry and knowledge production have, as a default, continued to place at its center of analysis what Audre Lorde theorized as the "mythical norm."[8] Defined by Lorde as "white, thin, male, young, heterosexual, christian and financially secure," this mythical norm occupies a position of privilege, or what Simone de Beauvoir called out as holding positions of both a positive and neutral subjectivity.[9] It is from this view of the mythical norm and the subject position of positivity and neutrality that current and dominant traditions of philosophy and science have developed their systems of seeing, naming, and defining the world.

In turn, the questions and answers derived from the ontological, epistemological, methodological, and ethical frameworks made possible within this dominant metaphysical tradition have been written into our textbooks. These frameworks have provided the language, ideas, and experiments for how we can come to know the world and how we are to orient ourselves toward the multitude of others that inhabit this world. As Claire Colebrook has stated, "the way we think, speak, desire and see the world is itself political; it produces relations, effects, and organizes our bodies."[10] If this is indeed the case, that how we see the world is political to the point of organizing our bodies, then something as simple as how we see and think about a blade of grass may matter a great deal to how we see and think about molecular biology, feminism, and where the two meet. If we want to reflect on how it is that we have come to know the world, we must also consider which questions have been asked about the bodies and biologies around and within us, why they have been asked, how they have been asked, and by whom. Much may be at stake in how we orient ourselves toward supposedly simple matters, and ultimately there could be much to gain if we follow through with the questions and answers that grow out of these encounters.

What Can Grass Do?

What can we learn to know if we begin our inquiries by thinking with the blade of grass? It may be of little to no surprise to agrostologists and serious gardener-types out there that in a Carl Linnaeus–inspired taxonomical world, grass has been classified as belonging to the kingdom Plantae, the division Anthophyta (flowering seed plants), the class Monocotyledons (embryos with one seed leaf), and if we keep going down the taxonomical ladder even further, into approximately 11,369 accepted different species according to the Kew Royal Botanic Gardens's online database aptly called GrassBase.[11] Although grass taxonomy, and the practices of naming varieties of grass and identifying the properties of what grass *is*, is itself a vast and productive field of analysis, this book focuses on a set of questions that can be applied not only to our knowledge of grass but also to feminism and molecular biology.

As the chapter epigraphs prompt us to ask, how does an organism such as a blade of grass happen? How does a blade of grass extend or retract itself through time and space? What does grass require in order to change, develop, grow, and perish? How does grass react to being touched and how does grass express desire? By turning to the abilities of grass, we can begin to think about matter and molecules not only through classificatory or representational terms but also through the question of *what they can do*. What can we learn from the abilities of creeping grasses to spread not only by developing new shoots that emerge from nodes on underground stems (rhizomes) but in some cases also by developing horizontal stems that grow above ground (stolons)? Posing this line of questioning is not only possible but has in fact become imminently relevant to the field of feminist STS and to a project such as the present one.

Rather than asking what counts as grass, defining what grass actually is only in terms of its chemical properties and physical structures, or using the traditional and dominant philosophical and scientific practices of naming, identifying, and placing various forms of grass into neat categories by pointing out what it lacks compared to other organisms and differentiating it from what it is not, this chapter takes a molecular approach to asking what matter, molecules, bodies, and organisms such as grass can do and are capable of doing. By taking a molecular approach to biological matter, bodies, and nature, I follow a line of questioning that is generally less explored in both feminist theory and molecular biology—one that pursues ideas of becoming more than being, movement more than stasis, change more than fixity, and intensities more than identities or essences in biology. This pursuit brings forward what might loosely be referred to as *biophilosophies of becoming*. This is not to say that there isn't any value to organizing our knowledge of grass and other biological matters along the principles of being, stasis, fixed subjectivities, stable identities. As scientists and philosophers, however, we are already familiar with the kinds of knowledge that can be gained by viewing the world through this dominant lens or more organized terms. We know, for example, that the chemical elements of grass include carbon, oxygen, nitrogen, phosphorus, and hydrogen, and that the enzyme chlorophyll enables the process of photosynthesis in grass. This knowledge is important, and the ontological frameworks and questions that motivate us to gain this knowledge are also important. There is no denying that scientific and social progress has been made and can continue to be made through the practices of naming and working with fixed identities.

By becoming a molecular biologist, however, I learned firsthand that arriving at a valid hypothesis requires a great deal of open-mindedness and, at times, what one might even call ontological flexibility.[12] One had to either learn to put aside their questions regarding the nature of being, or learn how to think about the nature of being and what one was bringing forth, as a direct result of their experimentation. By working with material actants such as genes, hormones, receptors, signal transduction proteins, bacteria, and *in vitro* cell cultures derived from transgenic mice, I came to question my own ontological assumptions and was ushered into orienting my thinking in terms of processes and becomings. Through my own research in molecular biology I came to appreciate what Natasha Myers has described as "excitable ontologies."[13] I am intrigued by the possibility of moving feminist politics and feminist STS projects even further into molecular modes of thought by developing biophilosophies of becoming that treat biology in terms of an event and molecular biology in terms of processes.

Why use the term "biophilosophy" and not "philosophy of biology" to describe this project at the intersections of molecular biology, feminism, and philosophy?[14] Although the philosophy of biology dates back to the early twentieth century and is well established as a discipline, Sahotra

Sarkar has argued, and I agree, that there has been a dearth of philosophical engagement with molecular biology itself. His own work is a rich genealogy of the missing history of philosophy in molecular biology, particularly the role of reductionism and the central dogma in molecular biology.[15] Sarkar suggests, however, that despite the discovery of the structure of DNA in the early 1950s, and the vast growth and impact of molecular biology research, "philosophical interest in molecular biology declined in the late 1970s and 1980s when, with few exceptions, philosophers of biology focused only on evolutionary biology and, within evolutionary biology, on the problem of identifying units of selection."[16]

Sarkar makes the case that by turning its attention to evolutionary theory, much of philosophy of biology is out of touch with contemporary biology, including molecular biology, let alone a field such as synthetic biology (discussed in chapter 5).[17] Instead, the philosophy of biology generally engages with a different set of questions. Paul Griffiths has summarized the three types of inquires that fall under philosophy of biology: addressing general theses in the philosophy of science through the context of biology, subjecting conceptual puzzles within biology to philosophical analysis, and "appeal[ing] to biology to support positions on traditional philosophical topics, such as ethics or epistemology."[18] Philosophy of biology has much to offer, but for the particular project I am developing here, traditional analytical approaches do not bring me immediately to questions I would like to pose.

Although there is a great deal of overlap, a distinction can be drawn between *philosophy of biology* and *biophilosophy*. Spyridon Koutroufinis has suggested that the borders between philosophy of biology and biophilosophy often shift, but some primary features can be used to distinguish the them. Whereas philosophy of biology turns to a materialism that stems from a mechanistic ontology and "metaphysical principles of classical physics, . . . in a version that is expanded to include the idea of dynamical systems, which include the theories of complexity, self-organization, and chaos," he suggests that contemporary biophilosophy holds a process-metaphysical perspective and takes into account "notions of matter and causality that have long been established within quantum physics."[19] Biophilosophy is also open to liberal naturalism, which according to Koutroufinis "allows mental states, such as phenomenal qualities, as aspects of natural entities and ascribes ontological relevance to abstract, modal, moral, and intentional entities."[20]

The biophilosophy I wish to explore here emerges at a moment within feminist STS, which due primarily to the work of Karen Barad can also be said to reflect a process-metaphysical perspective that engages with questions of matter and causality after quantum physics. It is also deeply influenced by feminist philosophical interrogations of subjectivity, found in the writings of Donna Haraway (whose work has been connected to Alfred North Whitehead's processual metaphysics), Elizabeth Grosz (whose work turns to Henri Bergson's emphasis on duration and time), and Rosi Braidotti (whose work elaborates Gilles Deleuze's ideas on nomadology).[21] These combined trajectories of feminist and materialist inquiry have led me to think more about feminist engagements with process ontology as well as recent interests in process ontologies for contemporary biology.[22] A feminist philosophy of biology has much to offer, but the biophilosophy that I am interested in articulating relies less on approaches of subjectivism found in feminist STS and more on those projects that frame the properties or qualities of becoming, movement, change, and intensities of matters and bodies in biology, in terms of process and events.[23]

Returning to the question of what grass can do allows us to think about organisms and molecules not as pre-given forms with fixed attributes, but instead, what the biophilosophy of Deleuze might encourage us to discern, as nonhuman becomings. Deleuze's biophilosophy is an ethical one, aimed at thinking beyond the human and, as some have argued, making way for a new metaphysics.[24] In our case, we may think of grass not as a fixed and passive entity, but rather as an event or what Deleuze and Guattari refer to as "haecceity"—that is, as a dynamic interaction between organisms and elements, one that experiences and expresses time and duration, has emergent properties, and is capable of change.[25] Drawing from the philosophical works of John Duns Scotus (1266–1308), Baruch Spinoza (1632–1677), and Henri Bergson (1859–1941), Deleuze and Guattari express *becoming* as that continual process of change and flux through which difference is produced.[26] This difference is positive and not defined through lack. It is through the production of a continuous difference that life simultaneously emerges, is sustained, and gets dissolved. Discussing the influence of Spinoza and Bergson on Deleuze's biophilosophy, Keith Ansell

Pearson explains that what Deleuze believes is that "what a body can do is never something fixed and determined but is always implicated in a 'creative evolution.'"[27] Ansell Pearson suggests that what Deleuze is faced with in thinking through difference and becoming is the development of a complex ontology, one in which we are "compelled to think an ethics of matter itself."[28] I return to this question of the ethics of matter more closely in the following chapters.

Resonating with Bharati, McClintock, and Tagore, in Deleuze and Guattari's *A Thousand Plateaus*, the authors use philosophies of becoming to motion us away from a dominant metaphysical tradition that has led us to believe that not only is there "a distinction between different orders of being" but that humans (albeit some more than others) are definitively and qualitatively separate from all other forms of life on Earth.[29] This tradition is marked by dualisms that work to separate "intelligibility and sensibility, doer and deed, Being and beings, condition and conditioned, [and] Creator and creation."[30] Perceiving our reality and existence in this way has given us the idea that humans are far superior to all other matters and forms in nature, and the supposed right to hold dominion over that nature. Bharati, McClintock, Tagore as well as Deleuze and Guattari suggest a different orientation, one that is both ontological and ethical. They present an alternate framework for perceiving our status as humans, and by doing so, force us to reevaluate our place in the world among nonhuman others. Instead of setting ourselves high above and far apart from the other organic and inorganic elements of our physical surroundings, they ask that we attempt to reduce this distance and get closer to such organisms as grass—an organism that grows under our feet and is quite literally "below" us. Rather than turning to a hierarchical chain of being marked by superiority and transcendence, the reorientation to the universe that Deleuze and Guattari promote is one of *univocity* and *immanence*. This reorientation requires moving away from the ideas of humans as ideal and autonomous subjects, and replacing our belief in a "great chain of being" with an understanding of the world that instead acknowledges the univocity of being.

The univocity of being put forward is not meant to suggest that everything is the same, or part of the "One" in some onto-theological sense; rather, as Brent Adkins has suggested, everything exists on a continuous and ontologically single field.[31] Deleuze and Guattari use the phrase "pluralism = monism" to make room for difference and the arrival of the new within this ontological univocity.[32] As Adkins explains:

> Dualisms create exclusive disjunctions, or biunivocal relations between terms. What Deleuze and Guattari are proposing is an *inclusive* disjunction by which we "arrive at the magic formula we all seek—PLURALISM = MONISM" (*Thousand Plateaus* 20). . . . The monism arrived at here, though, is not an Eleatic stasis in which movement is an illusion. It is the monism of the continuity thesis, the monism of univocity. The claim is not that ontology is a monotonous sameness, but that everything exists in exactly the same way. There is no dualism of form and content that must then be related by analogy. There is no transcendence, only immanence. All assemblages are arrayed on the same plane. The formula (pluralism = monism) is magic precisely because it allows for the creation of the new.[33]

Therefore, it should be understood that the *ontological univocity* of beings does not mean the erasure of difference. It also does not mean a dismissal of the distributive effects of power. This is a common misinterpretation of Deleuzian philosophy. Of course, we as humans are in many ways different from grass, and from mountains, rivers, and fields of corn, and for that matter from each other. However, is it possible to think about these differences without automatically assigning values on these differences? Are we as humans so very removed from other organic species and inorganic elements that we cannot see or feel a continuity with or connection to nonhuman others? The question perhaps is not whether or not differences exist between humans, animals, plants, water, inorganic elements, and more, but how we think about these differences. Do we choose to see these differences as positive differences, ones that are not understood through their lack or through their otherness from some transcendent figure or object? Do we choose to see these differences as differences in degree, or as differences in kind? By turning to the idea of ontological univocity, I am not interested in disregarding embodied experiences. Nor am I arguing for the equivalent but independent status of all objects through the lens of an object-oriented ontology or speculative realism. Rather, my intention is to see how the idea of ontological univocity can be used to emphasize our intimate moments of encounter with difference and to alter our treatments of these differences both inside and outside of the lab.

The remaining chapters are invested in exploring what biophilosophies of becoming can do for feminism, for molecular biology, and for the space where the two can meet. They deal with those theories, research designs, and techniques that are already present in both feminism and molecular biology that can help us to think about matter and bodies in terms of flux, motion, and capabilities. They consider more closely those approaches to knowing that can guide us in our encounters with other matters and bodies in the lab. The intention is to extend what becomes possible for both feminism and molecular biology if we think more about the capacities for change that exists in all aspects and expressions of matter, and the precise nature of our encounters with that matter.

I am not in any way invested in dismissing the knowledge that we have gained thus far, either through feminism or molecular biology, that relies on a logic of being, on a metaphysics belonging to the era of classical physics, or on mechanistic ontologies for that matter. I am not invested in discrediting the scientific method or scientific practices of gathering empirical evidence. I am, however, interested in exploring what new knowledges we can produce by thinking in terms of process and events through ontological univocity. More specifically, I want to consider carefully a biophilosophy of becoming that is highlighted by the qualities of (1) changefulness and nonhuman becomings, (2) kinship and hylozoism, and (3) univocity and immanence. These qualities are discussed in more detail below. In the chapters that follow, these qualities are further characterized through biological actants including bacteria, plasmids, *in vitro* cell lines, and minimal genome organisms.

Postcolonial and Decolonial Haecceities and the Project of Reframing

I am fully aware of the hesitation that may lie for many scholars in turning to Deleuzian concepts such as univocity and immanence, and particularly the fear of obscuring or "flattening out" identities, politics, and the effects of power that can come with the use of such concepts. This is why I am purposeful throughout *Molecular Feminisms* about placing these philosophical concepts into conversation with the ideas of anticolonial figures, and recent work in postcolonial and decolonial studies. The aim is to ensure some level of contextual accountability to this ontological and ethical framework. Of course, it must be understood that there is no singular or fixed context as such that can be said to define any given event. Just as matter, an event, or even the meaning of a text cannot be fixed, so too is true for context. Having said this, there is no point tiptoeing around my reason behind developing biophilosophies of becoming. Put simply, it is to learn how to deterritorialize or decolonize our thought by reframing dominant relations and practices found in both in feminism and in science.

Postcolonial, decolonial, and indigenous studies scholars have taught me the importance of considering the broader context of knowledge-making practices that come with a given philosophical concept or ontological gesture. They have taught me that one way to "decolonize relations and practices" is to give voice to a diverse range of knowledge bases in order to produce new ontological accounts.[34] In this way, concerns regarding the context of an event can also become opportunities to think about social justice. For instance, as decolonial frameworks, both feminist and postcolonial STS argue that "Western modern technosciences tend to distribute their benefits primarily to already well-resourced groups and their costs to economically and politically vulnerable groups."[35] Both fields also prioritize the concerns of those who have been marginalized or are considered "other" by the global North, calling attention to issues involving, but not limited to, the "environment, development, corporatization, and militarism."[36]

Sandra Harding explains that decolonial perspectives simultaneously allow for the "disunity of science" as well as "scientific pluralism."[37] In the case of Latin America, she states, "Latin American decolonial theory has been shaped by liberation theology, dependency theory, Paulo Freire's work, the distinctive history in Latin America (LA) of development in the context of persisting underdevelopment, and by chaotic recent economic and political histories in many of these countries. Different national histories have included different practices of inequality."[38] She writes that decolonial theory emphasizes the importance of "knowledge that is otherwise," where the term "otherwise" is understood as "an alternative to both neoliberal and Marxian understandings of democracy, anti-colonialism, modernity, tradition, capitalism, ontology, epistemology, and positivism."[39]

Similarly, Laura Foster sees the decolonial approach as "a set of research processes (and political practices) that seek to change the hegemonic ordering of knowledge production."[40] In her work on the hoodia plant, patents, and indigenous knowledge in South Africa, for instance, Foster sees the decolonial perspective as an important tool in the "project of reframing."[41]

Therefore, although some may consider it odd to bring Deleuzian thought together with postcolonial and decolonial perspectives to develop biophilosophies of becoming, my collaboration with postcolonial and decolonial STS scholars over the years has taught me several important lessons regarding the benefits of developing interdisciplinary encounters. Simone Bignall and Paul Patton, for instance, remind us that in her famous essay "Can the Subaltern Speak?" Gayatri Chakravorty Spivak points to Deleuze (as well as Foucault) as "being guilty of a Eurocentrism that fails to acknowledge how such 'speech' must be presented within the privileged structures of Western epistemology and representation in order to be comprehended or perceived as sensible."[42] Yet many postcolonial scholars believe that there is still some value to putting postcolonial theory into conversation with Deleuzian thought. For instance, Bignall and Patton attempt to draw some parallels between Deleuze's philosophical work and themes in postcolonial studies, including shared "comments about the imperialism of normative Western forms of Oedipal subjectivity; movements of de/reterritorialization describing a conceptual politics of capture and relative liberation; creation of hybrid and migratory forms of selfhood through relational processes of becoming, and of course [the idea of] nomads and their relation to the 'war-machines' that embody acts of resistance against the imperial 'state-form.'"[43] Closely related to the project at hand, Bignall and Patton see the value of Deleuzian concepts such as that of "minoritarian subjectivities and minor languages that introduce a deconstructive 'stuttering' into majoritarian identities, discourses and literary forms" for postcolonial work.[44]

Similarly, in reference to Deleuze and Guattari's theories such as becoming and the body without organs (BwO) and their relationship to postcolonial theory and feminism, Sushmita Chatterjee explains that the "theory is often seen as extremely Eurocentric and elitist in celebrating the ability to play at will. What does 'dismantling the self' mean for postcolonial subjects? Couldn't it be another insidious power ruse to distort the agency of subjects already dismantled through the politics of colonialism?"[45] However, like Bignall and Patton, Chatterjee also comments on the possible utility of these concepts in creating interdisciplinary conversations. "While recognizing these shortcomings," she writes, "it is also important to discern Deleuze's utility for postcolonial studies where this theory can be of great help to conceptualize minoritarian becomings, think about imaginative possibilities, and move beyond different limitations that colonize worlds and lives. . . . For postcolonial feminism, the 'politics of becoming' can work actively to decolonize relations and practices."[46]

The imaginative possibilities that Deleuzian thought brings to postcolonial theory (and vice versa, I suggest) is highlighted further by Rey Chow. Resonating with my own reason behind turning to Deleuze to work toward a biology that feminists desire, in a recent collection of essays Chow suggests that "following Deleuze's lead" would be "eminently logical for scholars to embark on an affirmative postcolonial studies, one that is less anxiously preoccupied with the mechanisms and apparatuses of European exclusion, perhaps, and more substantively engaged with the transformative potential" that concepts such as "becoming, deterritorialization, assemblages, multiplicities, affects, virtualities, bodies without organs, nomads, the rhizome, and so forth" have to offer.[47] Using what she calls "Deleuze's method," Chow admits to a certain utopianism in Deleuze's work, but she also suggests that this utopianism can be used by postcolonial studies to inspire liberatory thought. Thus, as I put forward biophilosophies of becoming for feminism and molecular biology, it is crucial for me that this philosophical approach attend to questions of context and work to decolonize certain established relations and practices. As I motion us toward molecular feminisms and biophilosophies of becoming, and ask us to consider biological organisms in terms of events, the question to remember is, How do we also consider the context of any such given event?

It is precisely the context of an event that Deleuze and Guattari are referring to by using the concept of "haecceity." This concept attends to the specificity and individuality of any given event. "There is a mode of individuation very different from that of a person, subject, thing, or substance," they state. "We reserve the name haecceity for it. A season, a winter, a summer, an hour, a date have a perfect individuality lacking nothing, even though this individuality is different from that of a thing or a subject. They are haecceities in the sense that they consist entirely of relations of movement and rest between molecules or particles, capacities to affect and be affected."[48] One way to understand a haecceity, and what makes any given event unique, is to think about things in terms of their ratios of motion and rest (what Deleuze and Guattari call longitude) as well as their intensities and affective qualities (what Deleuze and Guattari call latitude). The vast multiplicities of possible latitudes and longitudes contribute to the singularity of an event.

While discussing the multiplicity of how Deleuze's name itself has been taken up by so many postcolonial scholars, Réda Bensmaïa suggests that a haecceity allows us to ask specific questions about the events we are interested in analyzing such as, "What century are we in? What wave is sweeping us along? What history?" and "What new visibilities are possible after the postcolony?"[49] Bensmaïa makes the case that the concept of the haecceity helps us to simultaneously identify and name the "singularities which characterize forces, events, movements and moving objects, winds and typhoons" but also equally name a "period of time."[50] In this way we can come to see that the postcolonial as well as the decolonial are not only haecceities in and of themselves in the sense that they name a period of time, but also strategic approaches that we can use to identify and name some specific singularities that define an event. Postcolonial and decolonial haecceities make it possible to think about events in relation to specific dominant practices of knowledge production, particularly in relation to institutions that have in the past, and in some cases, still continue to support imperialism and colonialism.

For instance, in relation to postcolonial and decolonial STS, Itty Abraham suggests that "postcolonial techno-science[,] as a field of enquiry that crosses geopolitical boundaries as it tracks flows, circuits of scientists, knowledges, machines, and techniques[,] is a critical way of thinking about science and technology."[51] Abraham points out, however, that the emphasis that has been placed on the situatedness or the "local" within postcolonial STS (and within feminist STS, I argue) is often misunderstood. "When the postcolonial as a mode of analysis is linked to a fixed site of irreducible knowledge claims," he states, "it articulates an ontology that ties knowledge to location as a singular and essential quality of place."[52] Therefore, rather than using the postcolonial or the decolonial to mean the study of institutions and knowledge-making practices belonging to a specific location that was once or is still colonized, I use the terms more broadly. Such scholars as Itty Abraham, Suman Seth, Warwick Anderson, Sandra Harding, Vandana Shiva, Kavita Philip, Michelle Murphy, Gabrielle Hecht, Banu Subramaniam, Laura Foster, Kim TallBear, Anne Pollock, Amit Prasad, Kaushik Sunder Rajan, and more have shown that incorporating a postcolonial, decolonial, feminist, or indigenous STS analysis means reflecting on the situatedness and specificity of technology-mediated events, including biological events, in terms of capital, labor, time, geography, and scale—all of which can be examined through social institutions.[53]

Although the following is by no means a comprehensive list, recent work in postcolonial and decolonial STS emphasizes that scientific practices and technological interventions should be contextualized in a various number of ways, including an analysis of (1) transnational processes of colonialism and imperialism; (2) capitalist practices of production, consumption, and commodification; (3) gendered and raced labor of production and reproduction and the abstraction of this labor; (4) neoliberal forms of individualism and imperialism; and (5) technological impacts on global as well as local scales. Each of these approaches keeps an eye on the different elements that contribute to the specificity of an event while also developing a broad scope of analysis. These approaches have taught me to remember the situatedness of any given becoming and to remember that our understanding of biological and technological events is always connected to specific practices of knowledge production.

Changefulness and Nonhuman Becomings

Baba Bharati's epigraph at the beginning of the chapter speaks of the journey a molecule must initially make by becoming a blade of grass in order to find its place back to love. Here, what Bharati means by "love" is in fact the Lord Krishna himself, from whose bosom it is believed the universe was formed and to whose bosom all matter and forms are longing to return. Bharati (1868–1914) was a Hindu missionary who came to the United States to spread Krishna Chaitanya consciousness and Gaudiya Vaishnava theology. The quote is derived from the text *Sree Krishna: The Lord of Love*, written at the turn of the twentieth century.[54] The intention of the book, written as a Hindu theistic text in English by a colonial subject of British India, was to share a Hindu story of the history of the universe, from its birth to its destruction, to an American audience.

It must be noted that some postmodern critiques of science and rationalism have recently been aligned with "Hindutva" or renewed Hindu nationalist movements that pay increased attention to "Vedic science" to establish Hindu moral superiority.[55] I highlight the work of Bharati here to show that the idea of ontological univocity, which works against any sense of moral superiority, is not unique to the "West" or to continental philosophy alone. In the introduction to Bharati's text, he explains that the word "Krishna" in the ancient language of Sanskrit is derived from the root word "karsha," which means "to draw." Bharati explains: "Krishna means that which draws us to Itself; and what in the world draws us all more powerfully than Love? It is the 'gravitation' of the modern scientist. It is the one source and substance of all magnetism, of all attraction; and when that love is absolutely pure, its power to draw is absolute, too."[56] It is not my intention to question what we know about gravity through modern science or simply replace what we have come to know about gravity through the practices of scientific experimentation or the disciplines of mathematics and physics with the term "love." I am interested in what Bharati describes as the "draw" of love, and want to suggest that the attraction, magnetism, and pull toward something that occurs due to any form of longing may also be characterized in terms of desire.

Interestingly, it is this draw or desire that, according to this text, produces motion or movement in matter. Bharati explains: "All matter is changeful—matter is nothing but collected forms of change. Its seeming substance embodies but motion of change, so that its inmost attribute is changefulness."[57] Due to this quality of changefulness, he notes, nothing that is living in the universe is finite, "not even a blade of grass, or the tiniest speck of earth."[58] Bharati draws here not only from Hindu theology but also from scientific research and publications by the biophysicist Sir Jagadish Chandra Bose (whose work is discussed in more detail in chapter 2). Around the same time as the publication of Bharati's text in 1904, Bose was conducting experiments on electromagnetic waves, radiowaves, and plant behavioral biology. His experiments troubled the distinctions between the living and nonliving and suggested that the capacity to respond to a stimulus was not only present in humans and animal tissues but also in plants, metals, and other "nonliving" entities.[59] Drawing on these findings, Bharati extended the definition of what counts as living and therefore what has the capacity to express desire, changefulness, and movement, far beyond the human.

While explaining Deleuze's philosophy of becoming, Colebrook suggests that "becoming inhuman" is to understand that "life is the potential to differ"—a sentiment that fits nicely with Bharati's claim that "all matter is changeful."[60] "Becoming inhuman" resonates with Bharati's idea that in order to think about the history of the universe and the passing of time in new ways, we can think with the different temporalities and rhythms of grass. In fact, Colebrook argues that while developing a philosophy of becoming, Deleuze places a great deal of importance on different rhythms of temporality and scales of duration that are to be found in inhuman or nonhuman becomings. Colebrook states: "Deleuze seeks to expose an inhuman time that will open thought up to a future, a future that is no longer grounded on the unfolding of human history. . . . The history that Deleuze and Guattari compose in *A Thousand Plateaus* places human becoming alongside other planes of becoming. Within human life there are flows of varying speed and slowness—varying degrees of habit, memory, promising and desiring—while there are also the speeds and flows of non-human becomings (including animals, machines, molecules and languages)."[61]

In recent feminist and posthumanist projects, turning toward the nonhuman and decentering the human has been an important aspect of the critiques of anthropocentrism and the liberal humanist subject. In her most recent work, Elizabeth Grosz speaks to the role of the nonhuman and inhuman in our human endeavors. She writes:

> Art, science, and technology are not frames we impose on matter and ideality but explorations and inventions through the framing that incorporeals provide for our ongoing explorations of matter. They are contingent, contested elaborations of the world's qualities and processes. Art, science, and technology, not to mention the creation of economic and political systems, do not impose themselves from the outside on brute matter . . . but are rather the elaborations, in potentially infinite directions, of trajectories, lines of development, that are already there, immanent, in the prehuman and nonhuman world. It is to the prehuman, the inhuman, the organic and the inorganic, that we

> must direct our efforts, and which provide us with human ways to invent, to create ourselves and what comes beyond us.[62]

What is important here, and what can be thought of as the first set of tendencies that contribute to a project on biophilosophies of becoming, is to realize that becomings, and the capacity for changefulness, should be extended beyond human bodies to other organic organisms such as animals and plants, and even to inorganic compounds, molecules, and matter itself.

Kinship and Hylozoism

Barbara McClintock's epigraph would also have us reconsider our orientation toward grass. I am particularly struck by her sentiment that grass has the capability to scream.[63] McClintock (1902–1992) was a biologist who won the Nobel Prize in Physiology or Medicine in 1983 for her discovery of transposons or "jumping genes". She dedicated her life to studying the chromosomes of different species of maize starting in the 1930s and pioneered the field of cytogenetics.[64] McClintock biographer Evelyn Fox Keller has explained that developing a "feeling for the organism" was apparently a common refrain for McClintock while describing her own approach to scientific research. For me, this phrase of McClintock's has served as a refrain in my own efforts to develop feminist practices in the natural sciences that heighten our awareness of ethical engagements with what it is that we wish to know. McClintock's "feeling for the organism" returns us to the question of developing an ethics of matter, the broader implications of which are discussed in more detail in chapter 2. For now, however, I want to hone in on two points McClintock raises—namely, the ideas of kinship and hylozoism.

Interestingly, Keller downplays McClintock's claim that grass has the capability to scream, describing it as "an uncharacteristic lapse into hyperbole."[65] In specific reference to McClintock's statement that she feels sorry for walking on grass, Keller states, "a bit of poetic license, perhaps, but McClintock is not a poet; she is a scientist."[66] McClintock was obviously very attentive to the capabilities of nonhuman actants, and this willingness to describe the capabilities of grass using expressions commonly reserved for humans (such as screaming) should not be dismissed as crude anthropomorphizing. I would argue that McClintock wants us to reorient ourselves in such a way so that we may learn to be with grass, and to listen to what grass has to say, sing, and scream despite our physical, emotional, intellectual differences or presumed distance as species. She wants us to be aware of the fact that we do harm to grass, but I don't think that this awareness is geared toward preventing us from walking on grass altogether. Rather, it is beckoning us to recognize a sense of kinship or *co-becoming*.

Donna Haraway has spent a great deal of time and effort thinking and writing about a similar sense of kinship, whether through her work on cyborgs or dogs or most recently by drawing our attention to chthonic critters (those that dwell in the underworld).[67] "If there is to be a multispecies ecojustice, which can also embrace diverse human people," Haraway states, "it is high time that feminists exercise leadership in imagination, theory, and action to unravel the ties of both genealogy and kin, and kin and species. . . . We need to make kin sym-chthonically, sympoetically. Who and whatever we are, we need to make-with—become with, compose-with—the earth-bound."[68] A blade of grass may be about as earth-bound as it gets. It so happens that most species of grass grow either as rhizomes, with their roots joined in multiple networks just under the surface of the soil, or as stolons, with their roots growing as outwardly stretching veins running just along the surface of the ground. It turns out that making kin with grass, both sym-chthonically and sym-poetically, may not be such an odd place to begin after all.

However, McClintock's sentiment describes not only a sense of kinship between humans and grass but also a hylozoism that recognizes the capabilities of grass on an equal footing to those of humans. Hylozoism frames nonhuman forms, as well as matter that has yet to assume a form, as being active or "alive" in some way.[69] Rather than following the philosophical tradition of hylomorphism, where all matter is viewed as passive or inert until it assumes a pre-given form, through hylozoism one has to be willing to consider that all matter, even prior to the movement of this matter into any particular form or relation, has a self-sufficiency and the ability to exert some sort of push and pull on the universe.[70] For instance, the scientific discipline of taxonomy, which has named and divided grass into more than eleven thousand species, follows in the tradition of hylomorphism whereby clear distinctions are drawn between the subordinate properties of "raw" matter compared to those of actualized or pre-given forms. As useful as it is for organizational purposes, taxonomy is ultimately a practice of drawing lines between raw matters and forms. Taxonomy must go even further by separating forms from each other that are deemed as being different in kind. It is a scientific system that has been utilized to not only differentiate humans from their natural world but to give different elements, organisms, and even some humans a lesser or subordinate status along a supposed great chain of being. This scientific system requires us to deny the capacities for change that exists in all matter and to rule out the ontological reliance any given entity has upon another.

Lastly, I am struck by McClintock's claim that "basically everything is one," which in other terms may be referring to the univocity of being.[71] In my opinion, she is not trying to collapse, flatten, or disregard this difference. In fact, she spent her entire life analyzing the many different species of corn and learning the unique cytogenetic intricacies of each. This connection between the qualities of kinship and hylozoism led McClintock to develop a feeling for the kernels of corn that she studied; it is the ontological univocity articulated through these qualities that biophilosophies of becoming can bring forward.

Univocity and Immanence

Rabindranath Tagore's epigraph encourages us to frame our relationship with the blade of grass in terms of a live potential, thereby blurring the lines that have typically been used to divide humans from their nonsentient and nonliving surroundings. Tagore (1861–1941) was a Bengali polymath who won the Nobel Prize in Literature in 1913. A writer, poet, musician, and artist, he held a deep regard for even the smallest murmurings that could be found in nature. Tagore established Visha Bharati University in Santiniketan, India, to put into practice an educational philosophy and pedagogy that brought to the fore the importance of one's orientations toward nature. Tagore is noted for his profound sense of humanism, but this humanism was not the same as that humanism we think of today which holds at its center the modern liberal humanist subject. As Debashish Banerji has noted: "Tagore's critical humanism, rooted in a pre-Enlightenment Indian canon, included properties which exceeded the human, . . . we find a Tagore who, while including the freedom, justice, and poetry of the human, reached for identity beyond the human, a becoming-other, through affective empathy, an identification with existences beyond boundaries."[72]

Regardless of the religious or spiritual beliefs of Bharati, McClintock, or Tagore, we know that many organized religions such as Buddhism and Jainism would also have us consider the interconnectedness of the natural world more carefully. For example, in the case of Jainism, the principle of ahimsa provides its followers with a karmic impetus to do no harm to other life forms including animals, insects, plants, and microorganisms. However, all three thinkers highlighted here are pushing for an intimacy with grass that is not quite captured by a religious principle such as ahimsa. It seems that they are motioning us toward something other than a karmically-driven, nonviolent stance toward grass. Theirs is an ethics of encounter. Karen Barad has extended the onto-ethical relation to the ethico-onto-epistemological, marking the inseparability of ethics, ontology, and epistemology.[73] In later chapters we turn to the fact that none of these relations are without context, but for now let us remain within this onto-ethical plane. Given this starting point, the first onto-ethical maneuver we must deploy involves becoming more aware of changefulness and open to the capabilities of nonhuman others. Next, we must invite a sense of kinship and develop a more hylozoic view of the universe that recognizes the expression of certain capacities in all forms of matter. These capacities may not resemble our own, and of course these expressions will differ between animals, plants, and rocks. Recognizing these capacities brings us to the third set of qualities that contribute to biophilosophies of becoming: that of univocity and immanence.

The idea of ontological univocity would have us consider the differences that exist between humans and nonhuman others as existing on or even comprising an immanent plane of processes and becomings. In a traditional metaphysics characterized by hierarchical taxonomies and well-established orders of being, the gesture of bringing our human selves down to the level of an organism such as grass would undoubtedly be odd and may require knocking the illusions of our superior status down more than a notch or two. It is difficult from our vantage point as humans to summon such a sense of proximity to something as elemental as grass, let alone foster the ability to hear grass scream or acknowledge that grass can listen to music, unless we are prepared to learn how to live and play on a more level, more equal, or, in other words, an immanent field. The idea of ontological univocity levels this playing field, which in turn is an ethical maneuver. For Spinoza, and philosophers that have followed in his tradition including Deleuze, it is hard to keep a clear division between ontology and ethics. For these philosophers, ontology is ethics—and both exist on an immanent plane. In other words, how we think about ontology, or the nature of being, is in itself a matter of ethics, and in our case, how we can start thinking about an ethics of matter. Together, immanence and the univocity of beings allows us to imagine what it means to think about grass and other nonhuman and inorganic matters that sit in an audience hall *with us* and have the capacity to appreciate the songs and music produced not only by humans but also those songs created by clouds and forests alike. Paying attention to what grass or other organic and inorganic matters perceive is an example of a Deleuzian transcendental empiricism or as Colebrook has described, Deleuze's "inhuman philosophy."[74]

Becoming a blade of grass therefore is not about finding this organism's place along a hierarchical ladder or evolutionary tree, or highlighting what properties or abilities grass lacks compared to humans and other organisms. Neither is it about impersonating or mimicking grass. Rather, becoming a blade of grass is a biophilosophy of becoming that involves thinking about the qualities of changefulness and nonhuman becomings, kinship and hylozoism, as well as univocity and immanence. It is about making connections to organisms and elements such as grass so that we might find new ways to reach out and new ways to respond to the world around us. It is about using strategies, or what I refer to in chapter 2 as *microphysiologies of desire*, to approach difference in the world—not through lack but rather through positive and productive senses. Ultimately, it is about breaking our all too comfortable habits and opening ourselves to the molecular.

2

Microphysiologies of Desire

> I now proceed to a demonstration of the fact that whatever be the mechanism by which they are brought about, these plant responses are physiological in their character.
>
> —JAGADISH CHANDRA BOSE

> Compared to most animals, plant movement is slow; it occurs via growth processes and benefits accrue due to maintenance of connections between sister ramets. . . . Ramets remain connected via stolons or rhizomes for variable lengths of time and these connections allow for the transport of nutrients and hormones between the mother and daughter ramets.
>
> —ERICA WATERS AND MAXINE WATSON

Thanks to the entanglements of more than three decades of feminist critiques of science anchored in feminist health activism, feminist theory, and feminist philosophy of science, and more recent work taking place under the mantles of feminist materialisms and posthumanist ethics, there appears to be a particularly rich opportunity at this moment to think about feminism, molecular biology, and matter along more molecular lines of questioning. Why am I interested in biophilosophies of becoming that create minoritarian projects, deterritorializing gestures, and micropolitical sensibilities at the interstices of molecular biology and feminism? Biophilosophies of becoming can change how we perceive and encounter the world around us and, furthermore, can be used to support practice-oriented feminist STS approaches in the lab. As a feminist scientist, I have always been deeply curious about the relationships that form between the knower and what is to become the known. I have also been curious as to how, as feminist scientists, we are to proceed with our scientific inquiries once we start paying more attention to these relationships. Put differently, my time in the lab has made me curious about developing an *ethics of matter*.

I begin this chapter with a quote from an experimental account made by Jagadish Chandra Bose (1858–1937). Bose was a Bengali scientist who conducted his life's work as a colonial subject under British rule in India.[1] He pioneered investigations of radiowaves, microwaves, and even gave the first public demonstration of the existence of electromagnetic waves in 1895, managing to use electromagnetic waves to ring a faraway bell and even fire a pistol.[2] Bose, who can be described as the first biophysicist in our modern understanding of the term, used electrical signals to explore and trouble the limits of what we consider to be a life, or a living form. His scientific findings are fascinating, as are the instruments he designed to measure what he called "response" in animal tissues, plants, and metals. Two aspects of Bose's work spark particular interest for me as they relate to the project of using biophilosophies of becoming to decolonize and reframe dominant relations and modes of knowing in both feminism and science.

The first aspect is that despite having invented wireless telegraphy and the technology that led to the radio two years before Guglielmo Marconi took credit for the invention, Bose was adamantly against the idea of patenting and therefore chose not to lay such claims to ownership over his pioneering research. Instead, he was interested in the processes of discovery and apparently regarded the idea of patenting his scientific findings for monetary profit with a fair amount of disdain.[3] I am interested in this proprietary tension that people often raise when discussing Bose and his contributions to science, and what it might also have to say about Bose's alignment with a particular ontology of the natural world as well as his anticolonial stances within that world that might have informed his ethical relations to matter as a scientist. In addition to being a pioneering scientist, Bose belonged to the Brahmo Samaj, a Hindu reform movement that rejected polytheism, denounced the caste system, and played an important cultural role in the Bengal Renaissance. A key tenet of Brahmoism includes an understanding of god as being both infinite and singular, both immanent and transcendent, and one who is present in everything from fire and water, to plants and trees. Another key principle, specifically in reference to love, is that the reform movement asks its followers to respect all of creation.[4] It is quite possible that these principles also informed Bose's ethical approaches to studying and learning from the natural world.

The second aspect of Bose's work that I draw from involves his use of electrical activity as the measure of a *physiological response* in plants, metals, and animal tissue. His definition of the physiological properties of response and what constituted a "response" versus a simple "reaction" in nonhumans and nonorganic life were contested during his lifetime. More than a century later, scientists are returning to Bose's research to reevaluate the categorical distinctions we have drawn between humans, nonhumans, organic and inorganic life. Bose was the first scientist who convincingly argued that plants not only have a nervous system of their own but that they also have the ability to feel pain.[5] He demonstrated that the physiological ability to respond extends beyond the human to not only animals, plants, and microorganisms but even to rocks, metals, minerals, elements, and anything else capable of experiencing sensitivity to external stimuli. I am drawn to the ontology and ethics that Bose's approach presents and wish to use his claim of a physiology of response in this chapter to start reimagining our own feminist encounters with biology.

Despite many traditions of thought that have defined desire as a lack or as a negative concept, the abilities to respond, to act, and to react are also qualities that can be used to describe the notion of desire. Desire is what motivates us to reach out toward, grow closer to, and in some cases even touch the other. As Elizabeth Grosz has explained, drawing from Gilles Deleuze and Félix Guattari's attempts to provide alternatives to dominant psychoanalytic interpretations, desire can be understood otherwise. "Instead of understanding desire as a lack or a hole in being," she states, "desire is understood by Deleuze—again following Spinoza and Nietzsche—as immanent, as positive and productive, a fundamental full and creative relation. Desire is what produces, what makes things, forges connections, creates relations, produces machinic alignments . . . desire is an actualization, a series of practices, action, production, bringing together components, making machines, making reality."[6]

Bose's research on the physiology of response in plants helps me to better understand Deleuze and Guattari's idea of the micropolitics of desire, Grosz's work on feminist becomings, Donna Haraway's call to making kin, and Rosi Braidotti's posthumanist ethics. Following a transplanted line of flight from Bose's physiology of response, I put forward *microphysiologies of desire* as an *applied ethics of matter*. Microphysiologies of desire can be thought of as practices of encounter for feminists, feminist scientists, and scientist feminists.[7] They are a way to describe the naturecultural, as well as a strategy to proceed forward in our encounters with other humans, nonhuman animals, plants, bacteria, and inorganic others that comprise the naturecultural. Microphysiologies of desire serve as the material and semiotic nervous system of biophilosophies of becoming, extending beyond a single body and connecting the knower to what is to become the known. They put into practice and apply the ontological lessons drawn from biophilosophies of becoming by (1) cultivating an openness to nonhuman becomings and changefulness; (2) making connections through kinship and hylozoism; and (3) creating movement by way of univocity and immanence. Feminist philosophers and feminist STS scholars have been thinking about microphysiologies of desire in various ways, under various names, for decades. This chapter attempts to gather together these practices of encounter, not with the intention of creating one unified microphysiology of desire, but rather to show the rich and diverse ways that developing an ethics of matter has been at the heart of so many feminist desires.

Generative Desires

While revisiting her doctoral work on morning glories and color variation, Banu Subramaniam weaves together complex genealogies of evolutionary biology, eugenics, and invasive plants through feminist critiques of science, storytelling, and fiction.[8] Subramaniam poses the question, "How does one study the naturecultural world?"[9] She offers plant thigmatropism as a model mechanism, stating:

> An academy with separate and distinct disciplines has carved knowledge production into unique objects of studies and methodologies, obscuring the teeming life between the worlds of natures and cultures. . . . Inspired by the touch-sensitive thigmatropic tendrils of morning glories, which allow the plants to scale large objects and burrow into narrow crevices,

I narrate tales of morning glories through the curious and adventurous tendrils of naturecultural storytelling.[10]

Derived from the Greek root "thigma," which means "to touch," thigmatropism describes the ability of plants to sense and respond to changes in surface conditions when they come in contact with another object.[11] Subramaniam uses thigmatropism as her methodology for conducting interdisciplinary work. Inspired by similar tendrillic tendencies, but following more closely the immanent and stolonic extensions thrown out by crabgrass, this chapter develops a cartography for molecular feminisms.

My challenge in writing this chapter is to develop the project of molecular feminisms by learning from the capabilities of the stoloniferous plant. In the chapter epigraph, Erica Waters and Maxine Watson explain that stoloniferous plants such as strawberry plants grow both as a clonal colony (genet) that shares the same genetic material and also as individuals (ramets) that work together within this colony. Applying optimal foraging theory to plant growth, they explain that plants must be able to "sense, interpret, and respond to environmental signals."[12] For stoloniferous plants that grow by making aboveground connections in particular, apparently the "detection of differences in red/far-red ratios via phytochromes and other photoreceptors induces plant *morphological responses* such as enhanced elongation rates."[13] The morphological responses initiated by stolons therefore contribute to ramet growth across generations, from mother ramet to daughter ramet.

Accordingly, microphysiologies of desire aim to extend across and contribute to the growth of generations of feminist scholarship that have examined ontological and ethical approaches to matter. This does not mean that critical analysis of previous work becomes impossible or that tensions and disagreements between the work are overlooked. Rather a purposeful effort is made to avoid the tired and predictable Oedipal tendencies found throughout most scholarly traditions that would have us believe that theoretical progress can only occur by dismissing or discrediting previous forms of scholarship. Thus, instead of declaring the arrival of a "new" feminist approach to science, biology, the body, or a new ethics of matter by turning away from "older" work in feminist STS, feminist theory, or feminist materialisms, I use microphysiologies of desire to uncover lines of flight made possible by these previous engagements.

Similarly, Iris van der Tuin has articulated a need for moving away from our habits of characterizing a single and central lineage of feminist thought to thinking across generations. She advocates for generational feminisms with the suggestion that we use "jumping generations" as a feminist methodology. "The methodology of jumping generations," van der Tuin states, "changes the parameters of generational feminism and enables the abandonment of a feminist center, takes advantage of running on multiple and transversal tracks, and stimulates channeling one's energies and desires to seeking commonalities in difference and useful coalitions vis-à-vis current day problems. This is helpful for feminist politics in academia, art, and activism because it allows us to act on lessons learned from equality and difference feminisms when we discuss issues of representation."[14]

In my own research in molecular biology and reproductive neuroendocrinology, the act of bringing forward decades of feminist STS work on science, biologies, and matter through a generative lens has been crucial. These bodies of work directly motivated and challenged me to enter into a molecular biology and neuroscience wet lab in the first place. In the lab I participated in scientific projects where it was my responsibility to ask questions about the presence and mechanisms of androgen, estrogen, and melatonin receptors in gonadotropin releasing hormone (GnRH) neurons. I asked my scientific questions while keeping an eye to what it was that feminists wanted to know and change about *how* biological and reproductive physiology research was being conducted at the molecular level. For example, in *The Woman in the Body: A Cultural Analysis of Reproduction*, feminist anthropologist Emily Martin shared her research on the metaphors most commonly used in scientific and medical literatures to describe the processes of menstruation, birth, and menopause.[15] Martin provided evidence of gendered and hierarchical language being used to describe the cellular and molecular processes of reproduction mediated by the hypothalamic-pituitary-gonadal (HPG) axis in mammals. She argued that not only was this language being used to describe scientific results, but that the very design of scientific experiments on female and male reproductive physiology was influenced by these gendered and hierarchical paradigms. Her analysis, and the work of many other feminists invested in reproductive and social justice, led me to a lab where I had the opportunity to search for paradigm-shifting scientific evidence and create alternate modes of approaching the study of reproductive neuroendocrinology.[16]

Conducting scientific work in a lab was how I was able to practice my feminism, and I am very grateful for the generational support that led me there.

If you have ever watched stolonic grass grow, you know that there is no center from which the single blades emerge. Rather, runners crisscross on the surface of the soil, interrupting and integrating into already established patches. With the strategies of jumping generations and stolonic growth in mind, this chapter highlights those approaches that could be of most use to the feminist scientist who wishes to think about their encounters with the world through the scientific practices of biology. The idea of molecular feminisms itself is of course a direct take on difference feminisms. As I have stated, and as van der Tuin notes, difference feminisms and equality feminisms need not be at odds; they can work together. Just as a stolon can develop both horizontal shoots as well as vertical shoots that grow out of nodes, molecular projects are not in opposition to molar projects. In fact, van der Tuin's methodology of "jumping generations" draws out the generative capacity of thinking about the molar and molecular together through a generational lens. We can, for instance, connect the methodology of jumping generations itself to a rich genealogy of feminist philosophical work that includes Braidotti's theory of *transpositions* and Chela Sandoval's idea of *split consciousness*.[17] Both of these may in turn be connected to Barbara McClintock's scientific discovery of *transposons* or "jumping genes" in molecular genetics.[18]

Multiple generations of feminist materialist thinkers have made it possible for feminist scientists to consider questions of ontology and ethics more closely while seeing scientific research as a means of practicing their feminism. Following cues from such feminist STS scholars as Karen Barad, Patti Lather, Isabelle Stengers, and Haraway, and by drawing from the collaborative works of Deleuze and Guattari, as well as Keller and McClintock, the remainder of this chapter explores microphysiologies of desire that address the relationship between the knower and the to-be-known. Using the concepts of agential realism and intra-actions, getting lost, cosmopolitics, becoming-with, and a feeling for the organism, I present several different theoretical tools that feminists have already developed to think about an ethics of matter.

The chapter ends with my own idea of "feeling *around* for the organism," which I hope speaks to feminist scientists in the natural sciences and helps put together a shared vocabulary between feminism and molecular biology. What is perhaps slightly different about my argument, compared to past feminist STS projects, is my emphasis on the point that ontology and ethics are coetaneous and that learning *how to see the world* has always also been about learning *how to encounter* that world. My experiences of working directly with DNA, proteins, cells, cultured cell lines, bacteria, animals, chemicals, radiation, and machines in the lab taught me this crucial lesson about ontology and ethics. That is why while turning to the onto-ethical aspects of molecular feminisms, I focus on the importance of creating practice-oriented feminist STS approaches. I call upon those encounters that the feminist scientist must have with the everyday materials, tools, customs, languages, and theories of science. Reflecting upon these encounters makes it possible to think differently in the sciences and in many cases prompts us to pursue less explored lines of flight in feminist theory and politics.

Nonlinear Desires

Over the past several decades, the field of feminist STS has proliferated, producing rich feminist critiques of specific sciences and recuperating theories from the sciences for feminist ends. From the outset, feminists have pursued multiple theoretical approaches to thinking about science, ranging from "feminist theory of science," "feminist theory in science," "feminist theory out of science," and most recently "science out of feminist theory."[19] Feminists from different disciplinary and activist traditions have contributed *theoretically* to feminist STS, and as the editors of the special issue of the journal *differences* dedicated to "feminist theory out of science" suggest, everything we know and do is already "theory all the way down."[20] Undoubtedly, there are many ways to organize the impressive range of theoretical work that falls within the scope of feminist STS. Entangled in various productive ways *with* science, these intersecting but divergent feminist orientations have raised a host of questions about the nature of scientific knowledge production.

Distinguishing itself from metaphysics, religion, and superstition, "modern" science is often characterized as a knowledge-making practice that is rational, objective, empirical, experimental, and evidence-based.

Celebrating the capacity to discover the truth through systematic observation, hypothesis formulation, rigorous hypothesis testing, and falsification, the scientific method purportedly frees individual knowers from the taint of idiosyncrasy, bias, prejudice, particularity, and sociocultural values. As an interdisciplinary field that draws concepts and analytic categories from anthropology, cultural studies, economics, feminist theory, history, philosophy, political science, and sociology, feminist STS recognizes the importance and value of scientific inquiry but seeks to move beyond these characterizations to think about science as a social practice. Investigating how science works through its historical emergence, and the social, economic, and political dimensions of institutionalized science, feminist STS has illuminated how social values permeate the varied practices, processes, and products of scientific research. Building upon a core tenet of the field of science and technology studies, feminist STS has further illuminated the co-construction of science and society.

The field designations "feminist STS" and "feminist science studies" are often used interchangeably in the literature, and to complicate matters even further, feminist STS is also often referred to as "feminist technoscience studies." To many, these different designations have come to mean the same thing, and they do share a great deal of similarities in their disciplinary underpinnings and analytical frameworks. For example, all share common formative departure points such as critiques of positivism and pure objectivity. They all draw important theoretical insights into the body, biology, medicine and technology by reaching out to multiple sources of knowers and users. Scholars identifying with any one of these field names have responded to feminist, queer, intersex, and trans theory's interrogations of sex, gender, and difference. In fact, long aware of the limits of mainstream feminism's engagement with the concepts of sex and gender, both feminist science studies and feminist STS scholars have cultivated different sets of analytical tools. Also, rather than positioning race, class, sexuality, and disability as intersectional additives to a theoretical mainframe of sex and gender analysis, questions regarding the body, matter, materiality, difference, and nature in these fields have been articulated through much broader frameworks.[21] These frameworks have been attentive, as Murphy has summarized, to transnational processes of colonialism and postcolonialism, neoliberal capitalist practices of production, consumption, and commodification, and particularly in the US context, women's social justice movements that situated emerging reproductive and genetic technologies in relation to histories of slavery and eugenics.[22] Through a healthy feedback mechanism, many of these theoretical insights are now informing and reconfiguring key concepts that have typically been found in more canonized bodies of US- and European-based second-wave feminist thought.

However, the designations of feminist STS and feminist science studies indicate slightly different theoretical approaches and histories of disciplinary engagements that feminists have used to address questions regarding the role and impact of science and technology in our lives. From early work on the history of women in science, the persistent underrepresentation of women across STEM fields, and androcentrism in scientific discourses, the body of scholarship known as feminist science studies has paid particular attention to illuminating those practices in science that have led to the devaluation, marginalization, and exclusion of individuals based on such factors as gender, race, class, sexuality, disability, and colonialism. Rejecting the notion that these modes of exclusion are extraneous to science, feminist science studies scholars have sought to demonstrate that exclusionary practices are constitutive of particular scientific endeavors and that certain sciences have played crucial roles in consolidating constructions of women, people of color, the economic underclasses, the colonized, and the disabled as inferior and therefore less worthy of respect than elite property-owning white men. These studies have argued that in marked contrast to their claims of value-neutrality, various sciences have helped produce, sustain, and justify social inequalities and systems of domination. Much of this work was informed by commitments to creating socially just frameworks for conducting science.

Where feminist science studies can be noted for developing highly nuanced epistemological and methodological critiques of science, feminist STS scholars have drawn our attention to and emphasized the co-construction of science and society. Some feminist STS scholars have received their disciplinary training in the relatively new field of science and technology studies (STS) itself. STS draws primarily from historical and sociological studies of science and is mainly interested in delineating the relationships between scientific knowledge, technological systems, and society. Key contributions from actor-network theory developed in the sociology of science, from feminists scholars such as Lucy Suchman and Judy Wajcman who examined our relationships with technology, and from Haraway's figuration of the cyborg, helped form the field of *feminist STS* which has flourished into a larger umbrella term since the early 1990s.[23] With an emphasis on working with the sciences while also creating interdisciplinary dialogs, feminist STS has become a multidisciplinary field drawing from a various number of areas including feminist theory, black feminist theory, queer theory, disability studies, postcolonial studies, and STS (just to name a few). Commonly recognized analytical frameworks operational within feminist STS include (1) acknowledging the co-construction of science and society; (2) questioning the authority of science; (3) interrogating traditional definitions of scientific objectivity; (4) connecting feminist interventions in the sciences to community-based participatory projects and/or social justice movements; and (5) promoting practice-oriented approaches for knowledge production.

By highlighting these distinctions between feminist science studies and feminist STS, I want to be clear that it is not my intention to produce a narrative of increasing theoretical complexity or disciplinary progress. Nor is it my aim to dismiss the distinct interventions made possible by these approaches in an effort to produce one unified mode of inquiry. As Jutta Weber explained, regarding the development of the field of feminist technoscience over a decade ago, "the problem is how to write a non-linear and complex historiography of theories and practical engagements, as well as the artifacts of science and technology. It might help to avoid linear stories of feminist theory by reflecting not only on the epistemological and ontological framework of earlier approaches, but also by rethinking these frameworks in the light of contemporary sociopolitical developments as well as prevailing technological practices, artifacts, and material cultures."[24] In addition to reflecting on ontological and epistemological frameworks, developing new critiques of science and technology by pointing out essentialist assumptions, problematizing the use of binary categories, and questioning linear logic, some feminist STS scholars who are scientists use their training in the "hard" scientific disciplines to expose the "prevailing technological practices, artifacts, and material cultures" of science.[25] What I see as a standout feature of this scholarship is the attempt to develop interdisciplinary alliances and practice-oriented approaches in feminist STS.[26]

Posthumanist Desires

In this book I consider the capacities of life in the lab, ranging from bacteria, *in vitro* cell lines, and minimal genome organisms, to ask what it is that we as humans can learn from our exchanges with nonhuman actants. While thinking with these actants, as Weber suggests, I want to rethink previous feminist ontological and epistemological gestures in light of "contemporary sociopolitical developments" and "prevailing technological practices, artifacts, and material cultures."[27] The contemporary sociopolitical developments in which I am most interested include not only feminist, postcolonial, and decolonial projects but also posthumanist projects that aim to disrupt liberal humanist aspirations of autonomy and individualism.

The term "posthumanism" has come to describe several different schools of thought. The bodies of critical posthumanist work that I am referring to here are the discourses that developed through feminist theory, literary criticism, and cultural theory in the late 1990s. As Francesca Ferrando has suggested, this particular tradition of posthumanism brings with it an "awareness of the limits of previous anthropocentric and humanistic assumptions."[28] Describing key elements of this particular school of posthumanism, Ferrando states:

> Posthumanism is often defined as a post-humanism and a post-anthropocentrism: it is "post" to the concept of the human and to the historical occurrence of humanism, both based, as we have previously seen, on hierarchical social constructs and human-centric assumptions. Speciesism has turned into an integral aspect of the posthuman critical approach. The posthuman overcoming of human primacy, though, is not to be replaced with other types of primacies (such as the one of the machines). Posthumanism can be seen as a post-exclusivism: an empirical philosophy of mediation which offers a reconciliation of existence in its broadest significations.[29]

Although posthumanism allows me to think with nonhuman actants such as bacteria in order to question liberal humanistic narratives of productivity and progress made possible through autonomy and individualism, I do not wish to turn to this philosophical movement in order to ignore the human. It is not my intention to focus my inquiry on decontextualized artifacts of material cultures and thereby disembody these actants from their deep entanglements with human lives whose own conditions have been organized by systematic and institutionalized exclusions based on race, class, gender, and more. Haraway has criticized the term "posthumanism" on this very basis, claiming that it encourages the tendency of decontextualization. Instead, she has forwarded her concept of companion species as one that better captures the ontological and ethical entanglements that take place between humans and nonhumans.[30]

Explaining the importance of developing posthumanist ethics within gender studies, Cecilia Åsberg has turned to the field of animal studies. She motions to a body of work that has turned from anthropocentrism to the "integration of both human and non-human natures." Åsberg explains the need for developing such a "reciprocal ontology," drawing from both Haraway's companion species and Barad's concept of intra-action.[31] "Rather the ethical turn in this field [gender studies] is in the materialist wake of poststructuralist theory an attempt to recognize the other," Åsberg states. "Posthumanist ethics, entangled with onto-epistemologies of worldly 'intra-actions' (Barad), emerge as efforts to respect and meet well with, even extend care to, others while acknowledging that *we may not know* the other and what the best kind of care would be."[32]

My own experiences in molecular biology research have brought to light the need for such a reciprocal ontology, raising questions not only regarding the nature of existence and questions of being, becoming, and difference, but also how questions about our relationship with the physical and biological matters of the natural world can be articulated within the context of scientific inquiry. The reason for bringing together these critical discourses is that they have helped me, and I believe that they can help other feminist scientists who work with animals, cell cultures, or other nonhuman actants, to think through the ontological and ethical entanglements that occur at the level of the lab bench. For the feminist scientist, working in the lab matters. This in turn requires developing microphysiologies of desire that allow us to meet the other in the lab well, whomever or whatever that other may be, while acknowledging that we may *never completely know* that other.

Practice-Oriented Desires

In the 1989 pivotal feminist science studies anthology *Feminism and Science*, edited by Nancy Tuana, feminist scholars trained in philosophy, biology, and physics came together to discuss the intricate relationships between women, feminist theory, and science. Articles highlighted discordant views on the interventions in science that were made possible by feminist theory as well as the purpose and scope of feminist science studies (as it was referred to in the special two-volume issue of the journal of *Hypatia: A Journal of Feminist Philosophy*, from which the anthology was formed). In addition to providing an overview of feminist scholarship in the sciences at the time, Sue Rosser wrote in the anthology, "More feminists in science are needed to further explore science and its relationships to women and feminism in order to change traditional science to a feminist science."[33] Although other scholars in the same collection cautioned against the idea of a "feminist science" as such, Rosser's vision of changing traditional science to the promise of a feminist science provides a useful point of departure for thinking not only about the complex issues involved at the intersections of sex, gender, women, feminism, and science but also the need to develop practice-oriented approaches for those feminists who wanted to change the sciences from within.[34]

Rosser makes a clear distinction between the terms "women" and "feminism" that relates directly to my own argument regarding the project of thinking about feminist STS through both molar and molecular modes of politics. Although I am interested in developing a project in molecular feminisms, the argument I make does not aim to dismiss molar projects, such as those described by Rosser, that are aimed at exploring the absence or presence of women in science or "science and its relationships to women."[35] In fact, foregrounding the "women" question in science in epistemological terms has been very productive for the philosophical interrogation of science and knowledge production.[36] Rosser's own catalogue included feminist influences on pedagogical and curricular transformations in science, the history and professional status of women in science, feminist critiques of science, feminist theory of science, and even what she called the development of a "feminine" science.[37] Her additional call, however—the one that motions us toward exploring science and its relationship to feminism—is precisely where *microphysiologies of desire* can take us.

My interest in exploring microphysiologies of desire is directly related to the fact that I am indeed one of those "feminists in science" who, as Rosser challenged, did go ahead to "further explore science and its relationship to women and feminism."[38] By placing myself within the sciences, I faced both the challenge and opportunity of having my feminism tested and extended into less familiar areas of feminist thought. My experience has been stolonic in that my growth as a feminist STS scholar, much like crabgrass, has depended entirely on my ability to reach out and make intimate connections with less familiar modes of thinking. As a result of these connections, I have ended up here with my current molecular project. While this project is not opposed to liberal or equality feminisms that foreground "women in science" or "pipeline"-related questions, it does not immediately contribute to these inquiries. Rather, the promise of a "feminist science" for me has meant following the unexpected turns and outgrowths that result from the experience of becoming a feminist scientist. This path has led me directly to probe the relationship between feminism and science through minor literatures and less familiar means.

In an editorial written for the journal *Bioethical Inquiry*, Catherine Mills explained that the dearth of continental philosophy in the area of bioethics can be attributed to the fact that "recent continental philosophy has been more concerned with ontological questions than normative ones."[39] Arguing that the strict separation between questions of ontology and normative resolutions is neither correct nor useful, she also states that "continental philosophy is often criticized, if not derided, for a perceived failure to provide normatively clear guidelines about 'what should be done.'"[40] Although I agree with Mills that there need not be a strict distinction between ontological presuppositions and normative resolutions, I think that by placing a little more emphasis on the "what should be done" part of the equation, we can begin to invite more feminist scientists into this important conversation. Indeed, it would be incredibly useful for the feminist scientist in the natural sciences if a more concerted effort was made to connect interests in ontology, ethics, matter, and materiality with the everyday, nitty-gritty practices in the lab.

When a feminist scientist actually finds herself in front of a lab bench, she may be motivated to ask difficult questions that typically would not have been raised in her traditional scientific training. A likely place that her feminist research practices will first lead her are to questions that deeply interrogate the idea of who can be a knower and what can be known. More explicitly, how should she approach the object of study? How should she treat this object of study that is the other? What should she make of biological and statistical differences that emerge in the measurement of this other? How will these differences influence her understanding of subjectivity? Alternatively, the feminist scientist working in the lab may find that similar questions of ontological and ethical significance may begin to emerge as a direct result of her repetitive and ritualized performances of the scientific method. As a result of their attempts to know and to "discipline" the body, biology, and matter through scientific experimentation, many scientists (feminist or not) are dealing head-on with questions of ontology and ethics that are similar to those being raised within continental philosophy and feminist STS. With their hands-on experiences of working with live organisms and dealing with the difficulty of experimental reproducibility, biologists are accustomed to witnessing the fluidity, vulnerability, and unfixed "nature" of life.[41]

Adding to Mills's argument, I suggest that these ontological queries can in fact emerge from close encounters with the mundane or everyday techniques and tools that the feminist scientist requires in order to conduct experiments. In my own analyses of neuroscience, reproductive biology, and molecular biology research, for instance, I have been interested in how biological molecules and organisms are brought forward in the lab. Drawing from such theories as standpoint theory, situated knowledges, agential realism, and the methodology of the oppressed, I have called for feminist scientists who are working in the lab to examine the ideologies behind dominant representations of biological molecules and organisms and ask, Why not otherwise?[42] I have stressed the importance of bridging these ontological and ethical discussions with scientific practices. This is precisely "what can be done" in feminist STS.[43] As Michelle Murphy has reminded us in *Seizing the Means of Reproduction: Entanglements of Feminism, Health, and Technoscience*, multiple generations of feminist health practitioners and health advocates, starting in the late 1960s, have already used practice-oriented approaches to think about the body and biology differently. For example, by creating women's health clinics, designing their own tools for conducting vaginal exams, and producing pamphlets with anatomical details and methods for self-care, these feminists have devised their own ways of thinking about reproductive health and biology and went as far as to produce innovative forms of scientific knowledge.[44] Feminist health advocates show us every day that it is possible to simultaneously think with biology, work with matter, interpret data, and enact our feminist politics. They show us the importance of knowing what to critique and what to use from traditional scientific experiments and literatures. The scientific knowledge they have created continues to shape our understandings of female anatomy and reproductive health today.[45]

For the feminist scientist working in the natural sciences, ontological, ethical, and critical posthumanist concerns are always present, even if not clearly articulated as such. Ontological queries produced in the lab can very quickly become entangled with ethical queries related to one's research design. These entanglements might be traced back to the very beginning of one's inquiry, even before arriving at a hypothesis. Through the design of a research methodology, one must consider how to approach the encounter with what it is that one wishes to know. Driven by a molecular desire to position the knower in the same critical plane as that which becomes known, I will now flesh out a shared vocabulary for an ethics of matter that can be used in the laboratory setting of the natural sciences.

Indeterminacy in the Lab

As I outlined earlier, microphysiologies of desire are feminist practices of encounter. They help us articulate an applied ethics of matter and develop strategies for moving forward in our scientific work. Although we can begin to flesh out many such strategies, I would like to start with those that cultivate an openness to nonhuman becomings and the capacity for changefulness. In *Meeting the Universe Halfway*, Karen Barad advances ontological discussions in feminist theory by drawing from the physical and natural sciences. Through her knowledge of quantum physics, she invites us to reexamine and reformulate our current feminist theoretical treatments of matter and reality. In her introduction she shares the short story of an exchange between the quantum physicists Niels Bohr and Werner Heisenberg, leaving us with a powerful ontological lesson. Barad writes:

> For Bohr, what is at issue is *not* that we cannot *know* both the position and momentum of a particle simultaneously (as Heisenberg initially argued), but rather that particles do not *have* determinate values of position and momentum simultaneously. . . . In essence, Bohr is making a point about the nature of reality, not merely our knowledge of it. What he is doing is calling into question an entire tradition in the history of Western metaphysics: the belief that the world is populated with individual things with their own independent sets of determinate properties. The lesson that Bohr takes from quantum physics is very deep and profound: there aren't little things wandering aimlessly in the void that possess the complete set of properties that Newtonian physics assumes (e.g., position and momentum); rather, there is something fundamental about the nature of measurement interactions such that, given a particular measuring apparatus, certain properties become determinate, while others are specifically excluded. Which properties become determinate is not governed by the desires or will of the experimenter but rather by the specificity of the experimental apparatus.[46]

Extending this idea, we might start to recognize that the matters we study as biologists, for example, or that we attempt to define and then regulate, do not preexist. In fact, "we" as we tend to define ourselves as scientists and knowers also may not preexist but rather, as Barad suggests, participate in the "mutual constitution of entangled agencies." What becomes "determinate" or known is a result of the specific interactions of an apparatus. What constitutes the apparatus includes a range of players, including the knower, the tools of measurement, and discursive practices. Barad defines this mutual entanglement as an "intra-action." Intra-action addresses the question "What can we do?" Specifically, Barad would have us orient ourselves to an ontological and ethical framework that assumes indeterminacy and asks that we as scientists become accountable for "the material nature of practices and how they come to matter."[47]

Indeterminacy and accountability both play roles in the making of any phenomena and describes Barad's concept of agential realism. "In my agential realist account," she explains, "scientific practices do not reveal what is already there; rather, what is 'disclosed' is the effect of the intra-active engagements of our participation with/in and as part of the world's differential becoming. . . . What is made manifest through technoscientific practices is an expression of the objective existence of particular material phenomena. . . . Objectivity is a matter of accountability for what materializes, for what comes to be. It matters which cuts are enacted: different cuts enact different materialized becomings."[48] Barad's idea of the "agential cut" might be thought of as an event, or what I have described using Deleuze and Guattari's term: a haecceity. This cut allows for a "resolution of the ontological indeterminacy" and the "condition for the possibility of objectivity."[49]

To better illustrate the usefulness of Barad's feminist practice of encounter, I would like to share a very interesting case of indeterminacy accompanied by accountability in molecular biology research. A few years ago, Linda Buck stirred up a storm of controversy when she retracted the findings from one of her own groundbreaking scientific works published in the highly acclaimed scientific journal *Nature*. Embedded within this controversy were deeper questions related to issues of ontology, epistemology, ethics, and the nature of discursive practices. Buck, who studies the olfactory systems in mammals and the mechanisms involved in odor and pheromone sensing, shared the 2004 Nobel Prize in Physiology or Medicine with Richard Axel. Working as a postdoctoral fellow in Axel's lab in the 1980s, Buck successfully managed to identify a family of more than a thousand genes that code for odor receptors.[50] She has since spent a very productive scientific career mapping out the neurological and molecular basis of olfaction. Her work has revealed the interaction between olfaction and reproduction at the neuromolecular level.[51] Of her many scientific accomplishments, Buck and her colleagues are known for utilizing molecular visualization techniques such as genetic tracing methods to better understand the neural circuits involved in the regulation of the olfactory system. Using transneuronal tracers, Buck and colleagues have shown that gonadotropin-releasing hormone (GnRH) neurons "receive pheromone signals from both odor and pheromone relays in the brain" and that "feedback loops are evident whereby GnRH neurons could influence both odor and pheromone processing."[52] Her lab was the first to have engineered transgenic mice in which GnRH neurons also expressed the transneuronal tracer barley-lectin (BL) and green fluorescent protein (GFP). By performing immunostaining of brain sections derived from these mice, Buck has been able to visually map the neural circuits of GnRH neurons.

The controversy that emerged surrounding the scientific work involved Buck's research on the visualization of signaling from specific odorant receptors to specific clusters of neurons in the olfactory cortex.[53] In the March 2008 retraction of the original paper, Buck and her colleagues stated: "During efforts to replicate and extend this work, we have been unable to reproduce the reported findings. Moreover, we have found inconsistencies between some of the figures and data published in the paper and the original data. We have therefore lost confidence in the reported conclusions. We regret any adverse consequences that may have resulted from the paper's publication."[54] One of the reasons this retraction is so interesting and caused such a stir is that the retraction came from a Nobel Prize winner. Also, the retraction statement goes on to reveal that the actual experiments that were put into question were not done at the hands of Buck herself, but rather by one of the two primary authors of the article, who was a former postdoctoral fellow in her lab. In scientific circles the retraction of an article from a prestigious journal always makes for sensational news. Commenting on the retraction, the article "How to Read a Retraction" posted in the science blog *Drug Monkey* suggested how strange it was to see an "author contribution" list in the retraction statement—a list that did not exist in the original article and one that outlines the exact contributions of each scientist.[55] Basically, the "author contribution" information reveals that the postdoc's work was under question, and the postdoc was being made to take the fall for the faulty research.

How can this event be read? The question is not so much what Buck should have done differently, but what we can see and know differently as a result of this event. It is possible that the postdoc somehow fudged the results. From an agential realist account, however, we would also have to ask what effects or events have been disclosed as a result of the intra-active engagements of Buck, her other scientific colleagues, the retraction statement, and of course, the odorant receptors and neurons of the olfactory cortex.[56] It would appear that prior to November 2001, the intra-actions that had formed up until that point had resulted in a lack of scientific knowledge regarding the neural mechanisms of olfaction in mammals. In fact, until Buck's original work in the 1980s, the odorant receptors themselves (as we have come to know them) did not even exist. After November 2001, upon the publication of their findings in the journal *Nature*, Buck and colleagues, as well as the entire scientific community that supports a system of peer-review, faced yet another shift in the ontological status of these receptors. New intra-actions between scientists, mice, the transneuronal tracer barley-lectin, and visualization techniques disclosed a novel biological relation, one of signaling between olfactory receptors and neurons. After March 2008, however, this knowledge was once again put into question and a formal retraction published in a scientific journal, thereby in a way "dematerializing" that biological relationship which had come to be. How can we as scientists become more open to such nonhuman becomings and to such capacities for changefulness?

From a traditional perspective of scientific method and objectivity, it is hard to say what happened, and perhaps pointing a finger at the postdoc seemed like the easiest thing to do at the time. From the perspective of Barad's practice of encounter, however, we might suggest that the apparatus (that is, the combination of all the human, nonhuman, organic and inorganic actors and measuring devices that went into creating the examined phenomenon) changed and thus new agential cuts were enacted. Buck claims that she was no longer able to repeat the findings of this initial experiment in her lab. What is not readily known is that her lab moved from the time when the initial experiments were conducted. The results published in 2001 were based on work that her postdoc had done at the Howard Hughes Medical Institute at the Harvard Medical School in Boston. Buck tried to repeat the experiment, likely with a different postdoctoral fellow, a different generation of transgenic mice, and perhaps even a different water source to mix the chemical reagents needed for the experiment in her new lab in Seattle, at the Fred Hutchinson Cancer Research Center, University of Washington.

Following the theoretical insights of Barad's onto-epistemological framework, the "inconsistencies" that Buck refers to in the retraction statement can be read to imply much more than simply the faulty lab notes taken by a postdoc. Perhaps Buck's statement, in a way, also reveals the possibility for a new approach to dealing with biological matter. Through the appearance and disappearance of signals between olfactory receptors and specific neurons, we may be able to see a microphysiology of desire emerging from within the sciences—one that moves us from an ontology that treats what it encounters in biology as being fixed, to one of becoming that takes more seriously the ideas of fluidity, flux, and indeterminacy.

Becoming-With in the Lab

As some readers may anticipate, no chapter on developing feminist practices of encounter in the natural sciences would be complete without invoking the work of Donna Haraway. Her writing directs us toward those microphysiologies of desire that help us make connections through kinship and hylozoism. I am interested in bringing together and developing a vocabulary of ontological and ethical gestures that the feminist scientist might find useful in the lab. I am particularly interested in those microphysiologies of desire that help position the feminist scientist as a knower who operates in the same immanent plane as that which is to become the known. For some time, Haraway has turned her attention to the practices and effects of multispecies entanglements. In *Staying with the Trouble: Making Kin in the Chthulucene*, Haraway describes the significance of her concept of "staying with the trouble." She writes: "My multispecies storytelling is about recuperation in complex histories that are as full of dying as living, as full of endings, even genocides, as beginnings. In the face of unrelenting historically specific surplus suffering in companion species knottings, I am not interested in reconciliation or restoration, but I am deeply committed to the more modest possibilities of partial recuperation and getting on together. Call that staying with the trouble."[57]

In *When Species Meet*, Haraway developed this ongoing project on companion species relations. Throughout that book she expands on what thinking through companion species relationships could mean by asking two main questions: "(1) Whom and what do I touch when I touch my dog? and (2) How is 'becoming with' a practice of becoming worldly?"[58] Her interpretation of becoming-with begins with the idea that to respond is to show respect and that the practice of becoming-with works to "remove the fibers of the scientist's being."[59] To appreciate the idea of becoming-with as a feminist practice of encounter, Haraway paints a scenario of a scientist working within their discipline. Commenting on the work of primatologist Barbara Smuts, Haraway states:

> Trained in the conventions of objective science, Smuts had been advised to be as neutral as possible, to be like a rock, to be unavailable, so that eventually the baboons would go on about their business in nature as if data-collecting humankind were not present. Good scientists were those

> who, learning to be invisible themselves, could see the scene of nature close up, as if through a peephole. The scientists could query but not be queried. People could ask if baboons are or are not social subjects, or ask anything else for that matter, without any ontological risk either to themselves. . . . [I]f she really wanted to study something other than how human beings are in the way, if she was really interested in these baboons, Smuts had to enter into, not shun, a responsive relationship.[60]

Becoming-with informs how we as feminist scientists can start to think about experimentation, but goes one step further to disturb our ontological presuppositions of what in fact constitutes a knower and the to-be-known. As a relational ontology, becoming-with is clearly aligned with the concept of becoming. Haraway's work reads nicely together with a great deal of Deleuzian thinking. However, she is also quite vocal about distinguishing her concept of becoming-with from Deleuze and Guattari's use of becoming-animal, which she claims comes with a deep disdain for domesticated animals and, among other things, "incuriosity about animals."[61] Haraway places a different kind of emphasis on becoming by embedding it within companion species relations. "Becoming-with, not becoming, is the name of the game; becoming-with is how partners are, in Vinciane Despret's terms, rendered capable," she writes. "Ontologically heterogeneous partners become who and what they are in relational material-semiotic worlding. Natures, cultures, subjects, and objects do not preexist their intertwined worldings. Companion species are relentlessly becoming-with."[62]

"Worlding" is Haraway's way of communicating a particular kind of coming together, an enmeshment, or even a touch that becomes possible once we see life and all of its actants operating on an immanent plane. She explains:

> Instructed by Eva Hayward's fingery eyes, I remember that "becoming with" is "becoming worldly." *When Species Meet* strives to build attachment sites and tie sticky knots to bind intra-acting critters, including people, together in the kinds of response and regard that change the subject—and the object. Encounterings do not produce harmonious wholes, and smoothly preconstituted entities do not ever meet in the first place. Such things cannot touch, much less attach; there is no first place; and species, neither singular nor plural, demand another practice

> of reckoning. In the fashion of turtles (with their epibionts) on turtles all the way down, meetings make us who and what we are in the avid contact zones that are the world. Once "we" have met, we can never be "the same" again.[63]

Haraway's feminist practice of becoming-with involves giving up the idea of human exceptionalism. This is a key aspect of critical posthumanist projects as well. Haraway's ethical stance is guided quite emphatically by the desire and ability to touch and to be touched. It helps us work through the concepts of kinship and hylozoism and highlights an integral component of all microphysiologies of desire—namely, the ability to respond.

Getting Lost in the Lab

As one would perhaps anticipate, microphysiologies of desire can also create uncomfortable and complex encounters. They can motion us toward new lines of flight by way of univocity and immanence, but the outcomes are not always guaranteed to solve all of our problems. Yet through these difficult encounters, movement and change can occur. In *Getting Lost: Feminist Efforts toward a Double(d) Science*, Patti Lather theorizes "getting lost" as a feminist practice of encounter that also functions as a "fertile ontological space and ethical practice."[64] Lather's getting lost serves as a perfect example of how strict distinctions between ontology, ethics, epistemology, and methodology cannot easily be drawn. Lather articulates getting lost in the following way:

> At its heart, *Getting Lost* situates feminist methodology as a noninnocent arena in which to pursue questions of the conditions of science with/in the postmodern. Here we are disabused of much in articulating a place for science between an impossible certainty and an interminable deconstruction, a science of both reverence and mistrust, the science possible after our disappointments in science. Against tendencies toward the sort of successor regimes characteristic of what feminist philosopher of science, Sandra Harding (1991), terms triumphalist versions of science, this book asks how to keep feminist methodology open, alive, loose. . . . Given my interest in the science possible after the critique of science, my central argument is that there is plenty of future for feminist

> methodology if it can continue to put such "post" ideas to work in terms of what research means and does.[65]

Lather appreciates the importance of being able to work in a lab and continue to raise that pipette, despite our disappointments with science. This is indeed a crucial and challenging task for the feminist scientist. The list of disappointments is long and includes the distress caused by biological theories that have been used for deterministic ends and have contributed to the normalization of inequalities, the regret over biotechnologies that have caused environmental harm and have been produced at the cost of many lives, and the frustration that can come with positivism and the belief in pure objectivity. But obviously, if the feminist scientist is to continue in the lab, they must learn how to look beyond these disappointments and continue to navigate their steps—or as Lather puts it, learn to work within the "ruins."

The feminist scientist must learn to take the dilemmas and disappointments with existing technological practices and look at them in a different light. As Lather suggests in the plateau of her book dedicated to working within the ruins: "In such a time and place, terms understood as no longer fulfilling their promise do not become useless. On the contrary, their very failures become provisional grounds, and new uses are derived. . . . To situate inquiry as a ruin/rune is to foreground the limits and necessary misfirings of a project, problematizing the researcher as 'the one who knows.'"[66] Lather continues: "In this move, the concept of ruins is not about an epistemological skepticism taken to defeatist extremes, but rather about a working of repetition and the play of difference as the only ground we have in moving toward new practices."[67] Like Barad, in this brief statement Lather raises both ontological concerns as well as the ethical issue of accountability for "the one who knows." If the feminist scientist is to take these concerns to heart, how does one actually learn how to work within the "ruins" of their discipline? This is perhaps not the easiest task to undertake, particularly in the natural sciences. "Getting lost," as Lather suggests, is about becoming at ease with the idea of uneasiness. Learning to live with uneasiness is indeed most crucial here, a sentiment echoed in Haraway's call for "staying with the trouble."[68] Lather would have the feminist scientist see the benefits of getting lost by learning how to live without absolute knowledge and by respecting the demand for complexity.[69]

It may be of some benefit to look back and consider again the story of Linda Buck—neuroscientist, Nobel Laureate, and retractor of a published scientific work. Are we now able to see how microphysiologies of desire can be used to read Buck's scientific work in a new light? Can we learn to see how Buck herself may have "foregrounded the limits and necessary misfirings" of her project?[70] Can we learn to read her retraction and response to the entire affair as an attempt to be "accountable to complexity?" Is Buck (who may or may not self-identify as a feminist scientist) showing us the importance of "getting lost"? By getting lost, the feminist scientist is also able to move toward previously unexplored encounters.

We now have some sense as to how the feminist scientist may go about enacting new agential cuts even if they have to use the traditional scientific techniques and tools that are readily available. The point may not be to create "new or better" methods, but rather to work within the dominant tradition—in this case the scientific method—and gain what fresh knowledge they can from accepting the loss that accompanies the use of this method. The dilemmas that will occur by working with the traditional technoscientific practices and tools in the natural sciences should not become paralyzing and their disappointments should not stop the feminist scientist from continuing to stand in front of a lab bench. Rather, the movement that will occur from getting lost in this place and posing the question "How do I proceed?" may bring with it a new ethical orientation toward matter. Drawing from the work of Deleuze and Guattari, Lather explains that "big band theories of social change have not served women well. Here, something begins to take shape, perhaps some new 'line of flight' (Deleuze and Guattari, 1987) where we are not so sure of ourselves and where we see this not knowing as our best chance for a *different sort of doing* in the name of feminist methodology."[71]

As a microphysiology of desire, at first glance "getting lost" may be disorienting to the feminist scientist. We are used to following clearly labeled flow charts and neatly organized protocols. Getting lost may seem counterintuitive, but it does address the question, What can we do? By getting lost, Lather states that one can neither claim to produce better knowledge than her nonfeminist peers nor be chasing after the ultimate "truth." After years of spending time in a lab, I too am drawn to those feminist practices of encounter that are open to new lines of flight and encourage different sorts of doing. Working along a similar line of flight,

I now move the feminist scientist from a plateau filled with disappointments, wounds, and loss to one with a different set of productive discomforts. I turn to the work of Isabelle Stengers, who suggests that we need to learn how to take risks in scientific inquiry and search for moments of joint perplexity with other (more traditional) scientists. Both Lather and Stengers develop frameworks for inquiry that elicit sensory experiences, but whereas Lather's getting lost is more closely aligned with Derridean deconstruction, taking risks and searching for joint perplexities are microphysiologies of desire that move us in the direction of Deleuze's ontological univocity and immanence.

Cosmopolitics in the Lab

We have to be willing to acknowledge that feminist scientists aren't the only ones in the lab who have the capacity to be disappointed. Regular scientists also face disappointment. Following cues from Stengers and her "ecology of practices," I am interested in expanding upon the question of how a feminist scientist might be able to work with instead of against the science and perhaps the scientists that have produced these disappointments.[72] As Stengers might suggest in response to this query: "The problem for each practice is how to foster their own force, make present what causes practitioners to think and feel and act. But it is a problem which may also produce an experimental togetherness among practices, a dynamics of pragmatic learning of what works and how. This is the kind of active, fostering 'milieu' that practices need in order to be able to answer challenges and experiment changes, that is, to unfold their own force. This is a social technology any diplomatic practice demands and depends upon."[73] To me, this suggests that feminist scientists need to go deep into the methods and protocols of their research projects and gain an intimate knowledge of the inner workings of their experimental setups. This intimate knowledge will give them the tools they need to take a risk and start asking different questions. The quote also suggests that by going deep into the practices of their specific science, the feminist scientist will also be able to produce a different kind of encounter with the scientists around them.

In *Power and Invention: Situating Science*, Stengers explains the importance of taking "risks" in order to move forward with scientific inquiry.[74] In this microphysiology of desire or feminist practice of encounter, one has to take a risk in order to find those moments of experimental togetherness and joint perplexity that can be shared with other scientists. This concept of risk forms the basis of an ontological and ethical framework that Stengers refers to as cosmopolitics. In his foreword to Stengers's book, Bruno Latour expands the link between this notion of risk and Stengers's use of the term "cosmopolitics." He states, "There are constructions where neither the world nor the word, neither the cosmos nor the scientists take any risk. These are badly constructed propositions and should be weeded out of science and society. . . . On the other hand, there exist propositions where the world and the scientists are both at risk. Those are well constructed, that is, reality constructing, reality making, and they should be included in science and society; that is, they are CC [cosmopolitically correct], no matter how politically incorrect they may appear to be."[75]

Stengers's cosmopolitics adopts a notion of "risky constructivism," which, according to Latour, opens up what gets to count as scientific evidence in the first place.[76] This openness places cosmopolitics apart from those types of practices that either narrowly promote a kind of scientific imperialism or those that would dismiss the scientific method altogether. As Steven Shaviro has written in his scholarly blog *The Pinocchio Theory*: "She [Stengers] seeks, rather, through constructivism and the ecology of practices, to offer what might be called (following Deleuze) an entirely immanent critique, one that is situated within the very field of practices that it is seeking to change. . . . Stengers' vision, like Latour's, is radically democratic: science is not a transcending 'truth' but one of many 'interests' which constantly need to negotiate with one another. This can only happen if all the competing interests are taken seriously (not merely 'tolerated'), and actively able to intervene with and against one another."[77] The task for the feminist scientist is not to critique traditional scientific practices in order to dismiss them. Nor is it sufficient to simply learn how to tolerate such practices. Movement must be made from seeking to secure a position of transcendence and "truth" to one of immanent critique and "joint perplexity." This process may involve reorienting our encounters not only to biological matter and organisms in the lab but to the other scientists we find working around us.

Drawing upon the same quote from Latour, Sarah Kember, who works at the intersections of artificial life, biology, and cyberfeminism, suggests that "at the heart of Stengers's cosmopolitics is a philosophy in which scientific realism and social constructionism are not opposed."[78] This is similar to Barad's agential realism, which aims to move "beyond the well-worn debates that pit constructivism against realism."[79] However, Kember goes on to state that "Stengers advocates a philosophy in which the object, the thing, the world is recognized as having something to say for itself. It is about embracing the risk which is therefore posed to science and to the scientist."[80] This is perhaps where the microphysiologies of desire that we can draw from Barad and Stengers begin to part ways. Barad, for instance, would have us place an emphasis on the ontological, epistemological, and ethical implications of coming to know the world through phenomena, and also have us take account of our *responsibilities for* those mutually constituted others that we want to come to know. Stengers, while aware of the imbrications of human, nonhuman, and technological actors or actants in the world that we come to know through the sciences, motions us to also consider that the "thing in itself" has some sway in what comes to constitute an event, underscoring a different type of emphasis on relationality.

This ontological gesture made by Stengers occurs upon a more immanent plane and returns me to a point I raised earlier regarding the ontological presuppositions that have guided the majority of feminist STS inquiries into the relationship between the knower and the known. If, as Stengers believes, we are able to picture ourselves in a location of immanent critique, we might be able to see how microphysiologies of desire that rely on the concept of ontological univocity might guide our encounters between the knower and the to-be-known. This ontological stance shifts our ethical encounters toward the to-be-known slightly—from that of responsibility for the other to one of simply response. Following Stengers, we might say that in this plane of immanent critique and joint perplexity, all actants become partial knowers and that the real challenge is to learn how to respond to that other knower.

"Feeling Around" for the Organism

Having analyzed several feminist practices of encounter that explore the relationship between the knower and what is to become the known, I am interested in exploring how slightly different ontological stances might influence an ethics of matter. What kind of an ethics of matter might come forward when we move away from ontological gestures that emphasize responsibility toward the other, and move instead toward those gestures that simply require the recognition of a response?

To address this, I turn to what Barbara McClintock described as her approach to science, an approach based on developing a "feeling for the organism."[81] McClintock made this comment while being interviewed by Evelyn Fox Keller. It is not my intention here to attempt to channel McClintock or to get to the "real" meaning behind her statement. Rather, I would like to end this chapter by exploring what this statement can mean for developing microphysiologies of desire. At its basis, McClintock's "feeling for the organism" is all about the relationship between the knower and the known. What I want to pursue further here is the uncertainty that remains over the precise nature of that relationship. Where does the emphasis fall? Is it a feeling *for* the organism, where the emphasis falls on the *for* in a type of benevolent affection toward, or is it a *feeling* for the organism, where the emphasis falls on the *feeling*, in a manner of stolonic or tendril-like extension? The distinction I am trying to make is subtle, but our attraction and possible ontological commitment to one meaning over another will have an impact on an ethics of matter that follows. I suggest that one type of feeling for the organism describes a molar or transcendent mode of encounter, while the other carries a more molecular or immanent approach. In the first scenario the feminist scientist may learn to develop a feeling *for* the organism. Her interaction with that organism, an organism that is no longer seen as an object simply available at her disposal, is reevaluated so as to accommodate a new ethical relationship of responsibility between that scientist and the organism. This is how she becomes accountable. In attempting to develop a feeling *for* that organism, she will have to ask herself what her ethical commitment is toward that *other* organism.

For instance, in her 2012 essay "On Touching—The Inhuman That Therefore I Am" and her 2007 book *Meeting the Universe Halfway*, Barad turns to the philosophical work of Emmanuel Levinas.[82] Levinas's ethics is generally understood as a study of intersubjectivity aligned with transcendence, existence, and the human other.[83] Barad reads Levinas through diffractive means and draws upon posthumanist ethics to propose that as an epistemological-ontological-ethical framework, her concept of agential realism allows us to "turn our attention to our responsibilities not only for what we know but what may come to be."[84] She aims to reorient the relationship between the knower and what is to become the known, working toward an ethics of mattering. "We (but not only we humans) are always already responsible to others with whom or which we are entangled," she states, "not through conscious intent but through the various ontological entanglements that materiality entails. What is on the other side of the agential cut is not separate from us—agential separability is not individuation. Ethics is therefore not about right response to a radically exterior/ized other, but about responsibility and accountability for the lively relationalities of becoming of which we are a part."[85]

Barad makes clear that agential separability is not individuation and that our responsibility toward the other does not occur through conscious intent. She emphasizes the point that the other (nonhuman others included) is not seen as an exteriorized other, but rather as another within a relationship. Importantly, however, there is also a reconciliatory tone in the ethical response that Barad forwards, taking the form of accountability or responsibility toward that other—or what I argue might be interpreted as a molar feeling "for" that nonexteriorized other. Furthermore, this molar commitment may lend itself more easily to a mode of engagement that is moored in transcendence. As Grosz explains regarding ethics:

> Unlike Levinasian ethics, which is still modeled on a subject-to-object, self-to-other relation, the relation of a being respected in its autonomy and the other, as a necessarily independent autonomous being—the culmination and final flowering of a phenomenological notion of subject, Deleuze and Guattari in no way privilege the human, autonomous sovereign subject, or the independent other, and the bonds of communication and representation between them; they are concerned more with what psychoanalysis calls "partial objects," organs, processes, flows, which show no respect for the autonomy of the subject. Ethics is the sphere of judgments regarding the possibilities, and actuality of connections, arrangements, linkages, machines.[86]

Although I have emphasized the importance of recognizing that molar and molecular approaches always coexist, and in many cases need to work together, my interests in revisiting the idea of McClintock's feeling for the organism is more aligned with thinking about ethics through this second, more molecular sphere of possibilities.

From a parallel site of play I would like to toggle the switch and move from a sense of responsibility moored in transcendence to an immanent sense of desire which is in search of a response—any response. I want to move to the idea of immanence and envision a molecular "feeling around" for the organism. I return to Bose's claim that both the living and the nonliving are capable of response and extend this immanent capability into a microphysiology of desire. At the start of this chapter, I turned to Grosz for her treatment of desire as immanent, positive, and productive. She defined desire as that which forges relations and creates connections. "Desire does not take for itself a particular object whose attainment it requires," she notes, "rather, it aims at nothing in particular above and beyond its own proliferation or self-expansion: it assembles things out of singularities; and it breaks down things, assemblages, into their singularities. . . . As production, desire does not provide blueprints, models, ideals or goals. Rather it experiments, it makes, it is fundamentally aleatory; it is bricolage."[87]

Where biophilosophies of becoming and microphysiologies of desire take us is precisely to *an ethics of matter where desire experiments*. Feeling around for the organism serves as an applied ethics of matter that brings together the qualities of changefulness and nonhuman becomings, kinship and hylozoism, and univocity and immanence. As a microphysiology of desire, "feeling around" resembles a stolonic searching in motion, a reaching toward and touching of an always unfixed and incompletely knowable other, in search of a response—any response. These responses can be good or bad, full of living or dying, but in no way are they reconciliatory. Thinking about our encounters in the lab through these qualities, we begin to see that a "feeling for the organism" can also be a "feeling around" for the organism. What is at stake here in developing this feminist practice of encounter is that different types of inquiries, experiments, and lines of flight become possible in the lab, depending on which applied ethics of matter we follow.

It is hard to know where such encounters will take us. In fact, developing practice-oriented feminist STS approaches for the natural sciences can feel a little bit like putting a love letter into a bottle and sending it out to sea. It is hard to predict if a fellow feminist scientist working away quietly and diligently in a lab, without a connection to a broader feminist community, will ever find this letter, let alone pick it up and run with it. Over the past few years, however, I have been delighted to see the growing number of undergraduate double majors in women's studies and biology. It will no doubt be very exciting to watch as these feminists enter into labs and start feeling their way around.

3

Bacterial Lives

Sex, Gender, and the Lust for Writing

> Bacterial sex differs from ours relative to time. Bacterial sex, the first kind of sex on this planet, is speedy sex.
>
> —LYNN MARGULIS AND DORIAN SAGAN

> Language (word choice, metaphors, analogies, and naming practices chosen to explain scientific concepts) and visual representations (images, tables, and graphs chosen to illustrate scientific concepts) have the power to shape scientific practice, the questions asked, the results obtained, and the interpretations made. Rethinking language and visual representation in textbooks can help remove unconscious gender assumptions that restrict discovery and innovation, and thereby reduce gender inequalities. . . . In bacteriology, this includes removing scientifically unsound metaphors that present bacteria as sexed organisms.
>
> —*GENDERED INNOVATIONS*

Putting microphysiologies of desire into practice, this chapter discusses the kinds of molecular projects that become possible when we begin feeling around for an organism such as bacteria. It considers what new lines of flight can emerge at the intersections of molecular biology and feminism when we think with members of the domain bacteria and carefully reflect on the capabilities of these primordial, non-nucleated, unicellular organisms. Regardless of how we frame our scientific theories of evolution—whether through Charles Darwin's ideas on mutation and adaptation, or

Lynn Margulis's ideas on symbiosis—the qualities of changefulness and nonhuman becoming in bacteria are absolutely integral to how we think and what we come to know about genetics, DNA, and molecular biology today.[1] Recent interest in the human microbiome, for example, coupled with the growth of metagenomics, has fostered a great deal of scientific interest in the contributions made by microorganisms such as bacteria to our human lives. The Human Microbiome Project sponsored by the National Institutes of Health recognizes that microbes such as bacteria are essential to our health, providing essential vitamins, breaking down food to extract nutrients, bolstering the immune systems, and producing anti-inflammatory compounds that are needed to fight disease.[2]

We have several modern scientific methods at hand to measure the extent of our biological and genetic kinship with bacteria. However, putting this recent interest aside, our connection with bacteria should come as no surprise. Since the first human observation of microorganisms over three hundred years ago, those of us who have worked with them have long known the important lessons we can learn from organisms such as bacteria. In addition to playing a crucial role in the genesis of this planet and all of its inhabitants including humans, bacteria have a great deal to teach us not only about changefulness and nonhuman becomings but also about desire, response, experimentation, and communication through language, writing, and text. Through an analysis of bacteria, this chapter asks what new ontological, epistemological, and ethical lessons can bacteria teach us? In particular, I turn to a bacterium's ability to alter its genome; importantly, bacterial sex plays a crucial role.

The ability to conduct genetic engineering through recombinant DNA techniques is not even a human invention, but rather a skill set that we as human scientists have simply borrowed from our bacterial kin. By taking advantage of their capacity for changefulness, we have learned a great deal from bacteria, including the physical and chemical features of DNA, the complex processes that are involved in the transcription of genes into messenger RNA (mRNA), and the translation of RNA into proteins. Whether we realize it or not, almost all of the molecular biotechnologies that we use today are based on the bacterial ability to have sex through several mechanisms of gene transfer. As humans, we certainly need to develop a broader understanding of sex, and indeed much of this chapter is dedicated to that very project. By learning more from bacteria, we can begin to expand our understanding of sex to include the molecular politics and microprocesses of desire, response, experimentation, and communication.

The sex I refer to in this chapter, as well as the sex that Lynn Margulis refers to in the epigraph, is the intricate array of molecular mechanisms involved in the processes of gene transfer. Bacterial abilities to replicate and edit genomes, synthesize new proteins, and adapt their immune systems are at the heart of many recent molecular biology–based research and technologies today, including the industrial and environmental applications of synthetic biology as well as the biomedical and directed engineering applications of the Clustered Regularly Interspaced Short Palindromic Repeats (CRISPR)-Cas9 gene editing system. No genetic engineering technology, and perhaps not even the field of molecular biology itself, would exist without the spectacular capabilities of bacteria—particularly, the bacterial capability of having sex.

Margulis (1938–2011) dedicated her career as a biologist to thinking differently about evolution and to challenging dominant paradigms in microbiology. She fearlessly promoted a theory of evolution commonly referred to as endosymbiosis theory, symbiogenesis, or symbiosis, at a time when many of her scientific peers were simply not ready to listen. In a 1967 paper titled "On the Origin of Mitosing Cells," Margulis argued that eukaryotic cells originated from three fundamental organelles that were once separate and free-living prokaryotic cells—namely, mitochondria, photosynthetic plasmids, and the basal bodies of flagella.[3] Later, drawing from and promoting the work of the Russian botanists Konstantin Mereschkowski and Boris Kozo-Polyansky, Margulis provided microbiological evidence that similar symbiotic relationships between different organisms actually provided the driving force behind evolution and genetic variation.[4]

In her book *Symbiotic Planet*, Margulis explained that the bacterial capacity to change or transform through metabolic means is at the heart of the origins of life on earth. "The smartest cells," she wrote, "those of the tiniest bacteria, about one ten-millionth of a meter in diameter, continuously metabolize. This simply means they continuously undergo hundreds of chemical transformations. They are fully alive. Recent work has revealed that the tiniest, most simple bacteria are very much like us. They continuously metabolize, using the same components as we do: proteins, fats, vitamins, nucleic acids, sugars, and other carbohydrates. It is true that even the simplest bacterium is extremely complex. Yet its inner workings are still like those of larger life."[5] Just as Margulis turned to bacteria to provide an alternative paradigm to thinking about the origins of life on earth, she also turned to bacteria to reframe our conceptions about the origins of sex.

Collaborating with her son Dorion Sagan, Margulis's scientific theories on the origins of sex are beautifully complex and yet have been made entirely accessible to nonspecialists.[6] Although open to the possibility that sex may have evolved in "unstable molecules before the origins of life" before the ozone layer even formed around the earth, Margulis believed that sex began as a way for bacteria to repair their DNA that had been damaged by solar radiation.[7] Bacteria did this by using DNA that was introduced from outside of their own bodies.[8] A key argument throughout Margulis's work on the origins of sex is that as humans we need to develop a broader understanding of sex. For instance, Margulis emphasized the distinction that needs to be made between sex and reproduction, claiming that if we think with bacteria, and consider a more historical evolutionary perspective that actually decentralizes the human, we can see that sex is simply a "genetic mixing process that has nothing necessarily to do with reproduction as we know it in mammals."[9]

Margulis stressed that reproduction occurs when a cell or organism copies itself, which, depending on the species, can take place with or without the need for engaging in sex. "Everyone is interested in sex," she wrote. "But, from a scientific perspective, the word is all too often associated with reproduction, with sexual intercourse leading to childbirth. As we look over the evolutionary history of life, however, we see that sex is the formation of a genetically new individual. Sex is a genetic mixing process that has nothing necessarily to do with reproduction as we know it in mammals. Throughout evolutionary history a great many organisms offered and exchanged genes sexually without the sex ever leading to the cell or organism copying known as reproduction."[10] Making this distinction between sex and reproduction, Margulis simultaneously underscored the notion that sex should not be thought of as a unified and singular type of act but rather as a "multifaceted and widespread phenomenon"—an idea that we return to later in this chapter.[11] For now, I would like to highlight the rich intellectual curiosity that Margulis brought to the sciences of evolution, microbiology, and molecular biology. Framing sex as multifaceted phenomenon, she perceived gene transfer mechanisms including conjugation, lysogeny, phage transduction, transformation, and plasmid transfer as different types of bacterial sex acts.[12] Despite providing this elegant ontological shift in our understanding and grasp of sex, a tension emerges in her work.

Discussing bacterial sex, Margulis was at times surprisingly hesitant in her language describing some of the properties of sex that have come to be associated with gender. Her use of the terms "sex," "gender," "male," and "female" with respect to bacteria foreshadows the message in the chapter epigraph and explains the tone of the cautionary notice posted on the *Gendered Innovations* website regarding the use of "unsound scientific metaphors that present bacteria as sexed organisms."[13] Interestingly, while discussing the origins of sex and the mechanisms of bacterial sex, in many instances Margulis constructs the familiar scenario:

> A sexual being, by biologists' definition, has at least two parents; and "gender" refers to the differences between these two parents. If bacteria have "genders" they are very subtle. Conjugating bacteria, before conjugation, look and behave just like each other. During conjugation, though, the rounded form, the "male" bacterium with a "fertility factor" among "his" genes, injects DNA into a "female" recipient whose DNA lacks fertility factor genes. In this travesty of transvestism, the "female" owing to its possession of the fertility factor, now becomes "male." This genetic gift can be passed on, indefinitely, changing genders as it goes.[14]

With or without the scare quotes, and despite the use of the term "transvestism," Margulis and Sagan's explanation of bacterial sex moves our understanding of sex from not only being richly complex but also being rather complicated. In the time I have spent over the years feeling around for bacteria, I have learned that something that is complicated can be quite generative. In fact, the scholarly and disciplinary complications of learning about the fertility factor in bacteria have provided me with the motivation for getting lost and for learning how to work among the runes/ruins of molecular biology.

For instance, in the entire four years of my undergraduate education and training in microbiology, I learned about male bacterial cells that carried the fertility factor required for conjugation several times over.

However, the curriculum at my well-reputed institution never once mentioned Margulis's work. Not even in my course and lab work in bacteriology and parasitology, where we learned about the possible interrelationships that can form between organisms—including symbiosis, mutualism, commensalism, and parasitism—did I learn about Margulis's ideas on the evolutionary concept of symbiogenesis. Despite this knowledge gap or "ruin," I am forever grateful to my professors for sharing their passion for all things microbial. This passion and the accompanying sets of problematics led me to my encounter with viruses, and thereby to one of my first experiences as a budding feminist scientist. Microbiology taught me the value of thinking with scientific knowledges, languages, and practices that were not just complex but complicated. I didn't have a name for it then, but I stumbled across my first object of "joint perplexity" between feminists and scientists when I asked my virology professor if there was a difference in the etiology of HIV/AIDS in women compared to men.[15] Although the answer given was profoundly insufficient, this was my first attempt to bridge disparate fields of scholarship, to mix disciplinary vocabularies, and to learn not to steer away from either the complexities or the complications of asking and knowing otherwise. I'll admit that after years of conducting interdisciplinary work, I am drawn to those productive tensions that emerge when we as feminist scientists, scientist feminists, feminist theorists, feminist STS scholars, or those of us who wear a combination of these hats, turn toward and not away from such difficult questions.

Thinking alongside bacteria, this chapter deals with a series of complicated binaries. These binaries are often used to encounter and interpret the world around us, starting with the binary of sex/gender. I then move to the binaries of biology/culture and matter/language. These binaries have troubled, and have been troubled by, feminist scholars working in a number of different fields of expertise and, as it turns out, bacteria have played an important role in this work. Bacteria have served as key actors not only in molecular biology research and genetic engineering but also in the growth and development of feminist theory.

First the chapter looks at some specific outcomes of feminist projects that have utilized the sex/gender binary to advance molar projects such as the participation of women and marginalized groups in science. Although diversity in science, technology, engineering, and math (STEM) fields remains a key issue for feminism, the deployment of the sex/gender binary through molar projects that have been based on liberal/equal rights feminist frameworks in science have, for better or for worse, bolstered binarized sex differences research and produced new areas of scientific research including the field of "gender biology." After demonstrating how the sex/gender binary has been incorporated into scientific analysis, I discuss why some feminist scholars have attempted to move beyond a molar and binary view of sex and gender. Importantly, they have also attempted to completely reframe how it is that we think about the terms "sex" and "gender" in the first place and, as a result, how we think about other related binary oppositions.

Next, the chapter extends these reflections on sex and gender to the relationship between biology and culture as well as matter and language. It takes a generational feminist approach by reviewing key contributions made by feminist poststructuralist accounts of matter and the body, then turning to more recent feminist materialist critiques of these post-structuralist contributions. Recent feminist and materialist scholarship attempts to recover from the purported split between sex and gender that was promoted by second-wave feminism. This has produced important insights for some feminists, particularly those trained and housed in cultural theory, to make their way back to thinking about biology, matter, and language. After several decades of theorizing a firm distinction between sex (read as biology) and gender (read as culture), particularly in English-speaking feminisms, many feminists are now finding the analytical tools of scientific research and feminist STS particularly useful in navigating a path back to sex, biology, and matter. Many are making this return by paying close attention to the complexities of biological processes.

The chapter ends by charting this "material turn" as it has played out specifically in the context of bacteria. I attempt to think more closely with bacterial lives. How can *biophilosophies of becoming* and *microphysiologies of desire* help us to know bacteria and therefore the world within us and around us otherwise? How should new connections and kinships with bacteria be understood in an era where recent advances in molecular biology have led us to the creations of synthetic biology and to genome editing technologies such as CRISPR-Cas9? If we return to a crucial feature of molecular feminisms, we are reminded that all ontological and ethical gestures, including biophilosophies of becoming and microphysiologies of desire, must be held to some kind of contextual accountability, which is different from enacting a transcendent mode of responsibility. Therefore, within this new era of molecular technologies, I suggest that we see bacteria and bacterial skills as events that can be reframed and understood through postcolonial and decolonial haecceities. It is clear, for instance, that these molecular technologies rely on bacteria as raw sources of labor in the production of new forms of biocapital. Similar to feminist STS, postcolonial and decolonial STS frameworks are rooted in social justice epistemologies. These approaches insist that our relationships with the matter and materiality of organisms such as bacteria are not just peripheral to science but are in fact constitutive of the histories, presents, and futures of scientific knowledge production. They remind us how important it is to think through both the molar and the molecular.

More Than "Add Women and Stir"

During the 1970s and 1980s in the United States, women's rights activists mobilized to raise national awareness of gender discrimination in all aspects of life, public and private. Some of these activists documented and sought to redress the systematic underrepresentation of women in academic, business, political, religious, scientific, and technological careers. Motivated by calls for inclusion from the women's movement and from members of Congress, government agencies such as the National Science Foundation (NSF) began collecting data documenting the underrepresentation of women and minorities in STEM fields.[16] Within a decade the NSF moved from tracking the number of women and minorities working in STEM fields in 1991 to implementing a grants program in 2001. This program—named Increasing the Participation and Advancement of Women in Academic Science and Engineering Careers (or ADVANCE)—was primarily aimed to "develop a more diverse science and engineering workforce."[17] According to the goals of ADVANCE, science itself did not need to change; it simply required a more diverse workforce.

Proponents of more inclusive science noted not only the absence of women as practicing scientists but also their exclusion from clinical trials. Again, in response to activist agitation and a congressional mandate, in 1994 the National Institutes of Health (NIH) issued a policy and created specific guidelines for the inclusion of women and minorities as subjects in clinical research involving humans.[18] Although the premise was that science itself did not have to change, the "add women and stir" approach in clinical settings pressed scientists to rethink their assumption that the adult male body was the norm on the basis of which all other bodies should be measured. Consider, for example, the announcement made in 2013 by the Food and Drug Administration (FDA) that zolpidem, the active agent in many sleep aids including Ambien, is metabolized differently in women than in men and that in women there is an increased risk of "next-morning impairment for activities that require complete mental alertness, including driving."[19]

To offset that risk, the FDA recommended that the "dose of zolpidem for women should be lowered from 10 mg to 5 mg for immediate release products (Ambien, Edluar, and Zolpimist) and from 12.5 mg to 6.25 mg for extended-release products (Ambien CR)."[20] Although the biological and molecular mechanisms for this difference in zolpidem drug processing are still not known, it turns out that the FDA did know back in 1992 that there was, as they say, an "effect of gender" when they conducted the clinical trials of zolpidem.[21] However, since the NIH policy regarding the inclusion of women was not implemented until 1994, scientists working on zolpidem apparently did not know what to do with this information regarding the effect of gender. Had scientists included women in clinical trials from the get-go and decided to learn about the causes behind the effects of gender that they observed in pharmaceutical research, years of overdosage in women could have been prevented. As this makes clear, the absence or disregard of women from most clinical trials has been problematic, not only for the health of women but also for the adequacy of scientific assumptions and the validity of scientific findings. Despite growing evidence of the importance of including diverse populations in scientific research, the most basic lab science involving animal research continues to use only the cells, tissues, organs, and bodies derived from males.

To address this deficiency, in May 2014 the NIH unveiled a new policy to ensure that preclinical research was sex-balanced by including female animals and cell lines obtained from females.[22] For some feminist scholars, however, simply including more women or female animals in biomedical research is equally problematic.[23] For it is often not known whether different results, which appear to be related to sex or gender, might be caused by an intervening variable. In the zolpidem case, the difference may have more to do with variations in height or body weight rather than with sex alone. Many feminists have even questioned how it is that scientists think they can go about isolating sex as such within a body, when it is not at all clear whether sex is even a factor that has a discrete biological location.[24] For example, is sex a genetic property that is easily identified through one's chromosomes? Or is it a factor that is located in the cells, hormones, genitalia, tertiary physical characteristics, or some combination of these locations?[25] Thus, for many of these scholars, a presumption of sex difference may be as troubling for scientific investigation as the absence of diverse populations from scientific research.

In light of such vexing questions, in 2009 a research team led by Londa Schiebinger created the web-based resource, *Gendered Innovations in Science, Health and Medicine, Engineering and Environment*. Jointly sponsored by Stanford University, the NSF, and the European Commission, this resource seeks to transform the inclusion question from an additive model to a more sophisticated deployment of sex and gender as analytical categories in research design to produce new discoveries. According to *Gendered Innovations*, project goals are to develop "practical methods of sex and gender analysis for scientists and engineers" and to provide "case studies as concrete illustrations of how sex and gender analysis leads to innovation."[26] The project explicitly seeks to move researchers beyond thinking about the category "women" as a "subgroup" of scientists who can diversify the workforce or as an additional variable to be inserted into an existing experimental protocol. The site pays particular attention to presenting feminist epistemologies and methodologies in science in an accessible manner for the scientific expert who is not trained as, or inclined to identify as, a feminist researcher.

In addition to providing examples of practical methods and case studies, the project defines key terms, including sex and gender, introducing feminist concepts to scientists who are unfamiliar with feminist scholarship. In keeping with the sex/gender binary, sex is defined as "a biological quality" and gender as "a socio-cultural process." Although these terms are treated as analytically distinct, a section titled "Interactions between Sex and Gender" emphasizes that the terms are not independent of one another: "Sex and gender also interact in important and complex ways. Rarely does an observed difference between men and women involve only sex and not gender, and rarely does gender operate outside the context of sex. The precise nature of their interaction will vary depending on the research question and on other factors, such as socioeconomic status, or geographic location, interacting with sex and gender."[27]

Perhaps it is this binary distinction, yet an emphasis on interaction that causes scientists, including Margulis, to easily conflate the terms "sex" and "gender." To add to this confusion, there is also the complicated nature of their interaction. As the section "Analyzing How Sex and Gender Interact" on the *Gendered Innovations* website explains:

> "Sex" and "gender" are distinguished for analytical purposes. "Sex" refers to biological qualities, and "gender" refers to socio-cultural processes. In reality, sex and gender interact (mutually shape one another) to form individual bodies, cognitive abilities, and disease patterns, for example. Sex and gender also interact to shape the ways we engineer and design objects, buildings, cities, and infrastructures. Recognizing how gender shapes sex and how sex influences culture is critical to designing quality research. Sex and gender also intersect in important ways with a variety of other social factors, including age, socioeconomic status, ethnicity, geographical location, etc.[28]

The *Gendered Innovations* project represents a committed attempt to work with basic research and industry scientists and to have them seriously consider the epistemological moorings of their scientific research by rethinking research priorities, reformulating research questions, considering participatory research and design methods, rethinking language and visual representations, and more. The suggested revision of bacteriology textbooks to remove "scientifically unsound metaphors that present bacteria as sexed organisms" can be found in a subsection titled "Textbooks: Rethinking Language and Visual Representations."[29] The revision of textbooks is presented as a prime example of a gendered innovation. That this project exists and the website serves as a guide and reference for scientists is an important indication that feminism has indeed worked hard to change the sciences or, at the very least, has much to offer to change how we continue to do science into the future.[30]

This project represents progress. It goes a long way to move scientists beyond the "add women and stir" mind-set and introduces scientists to the idea of co-production through the categories of sex and gender. Yet we should draw attention to the possible consequences of a gendered innovation that draws such clear binary distinctions between sex and gender and that relies on an interactionist framework. The *Gendered Innovations* chapter epigraph states that bacteria should not be thought of as being sexed organisms. In light of recent work in feminist theory, material feminisms, and feminist STS scholarship, I wish to explore what else it might mean to say that bacteria are sexed organisms and what else bacteria might have to teach us so that we can think more robustly about sex.

Conjugating Sex and Gender

Not without good reason, the *Gendered Innovations* project relies on a clean ontological distinction between sex and gender. However, encouraging scientists to consider sex and gender as distinct categories, even if an interactionist framework is added to their analysis, creates a new set of quandaries. For instance, feminist, queer, and postcolonial critiques of "inclusion-politics" have stressed that although no one would want women or other minority groups to be excluded from biomedical and technological research, the use of binary and interactionist models of sex and gender in science and medicine could work to further essentialize these identities. Simultaneously, these models do little more than pay lip service to other crucial interacting factors such as race, ethnicity, class, ability, and sexuality.[31] Not to mention, it is also possible that the inclusion of a specific identity-based group such as an ethnic minority through "race-based medicine" may in fact make individuals who identify with this group into new potential targets for profit-motivated pharmaceutical companies.[32]

Turning to biology and physics for alternatives to such binary frameworks, feminist STS scholars have advanced an important critique of "interactionist" approaches. The problem is that rather than troubling the boundaries between sex and gender, interaction presupposes the existence of two separate realms, which then come into contact. The limits of an interactionist frame can be illustrated by emerging research in gender-based medicine or what has also been referred to as "gender biology."[33] In their attempts to investigate how gender interacts with biology, typically understood in terms of a gene or a group of genes that code for or regulate what have been deemed as masculine and feminine traits, scientists often envision a Venn diagram. In this Venn diagram, biology occupies one sector, gender occupies another sector, and the space of overlap signifies the prospect of potential interactions. Although this limited interactionist framework could be used to help illuminate how gender inequalities (or race or class for that matter) produce biological *effects* such as health disparities in disease distribution and mortality rates, scientists often tend to interpret the model in a far more essentialist mode. The study of gender biology has come to represent an examination of the interaction of biology and culture in a manner that is reminiscent of sociobiology. Relying upon an ontological framework that produces a fixed biology (whether in the form of gene expression or hormones secretion), biological matters once again become the starting point of explaining the *causes* of perceived gender differences.

Philosophers of biology and feminist biologists have theorized alternative models in order to avoid the trap of biological determinism that is so frequently associated with the sex/gender binary system and its accompanying interactionist paradigm. They have turned to developmental systems theory (DST), for example, which offers a framework for understanding biology and development in relation to several major factors, including (1) joint determination by multiple causes, (2) context sensitivity and contingency, (3) extended inheritance, (4) development as construction, (5) distributed control, and (6) evolution as construction.[34] DST conceptualizes organisms beyond familiar binaries such as nature/nurture, genes/environment, and biology/culture. Anne Fausto-Sterling has perhaps been the most vocal champion of this theoretical frame through her work on intersex issues, bones, and more recently gender development in infants.[35] Yet DST also tells a cautionary tale for those who seek to move beyond dualisms by turning to interactionist conclusions. As philosophers of biology Susan Oyama, Paul Griffiths, and Russell Gray have noted:

> The standard response to nature/nurture oppositions is the homily that nowadays everyone is an interactionist: All phenotypes are the joint product of genes and environment. According to one version of this conventional "interactionist" position, the real debate should not be about whether a particular trait is due to nature or nurture, but

> rather how much each "influences" the trait. The nature/nurture debate is thus allegedly resolved in a quantitative fashion. . . . DST rejects the attempt to partition causal responsibility for the formation of organisms into additive components. Such maneuvers do not resolve the nature/nurture debate; they continue it.[36]

In order to avoid this outcome, Karen Barad uses the concept of "intra-action" as a means to move beyond the binary concepts of sex/gender, biology/culture, and matter/language. Advancing what she refers to as an agential realist ontology, Barad suggests that "the primary ontological unit is not independent objects with independently determinate boundaries and properties but rather what Niels Bohr terms 'phenomena.'" As discussed in the previous chapter, Barad advocates this ontological view of phenomena in response to questions of social constructivism, materiality, and more specifically, to feminist and queer reworkings of sex and gender. Intra-action is key to this ontological framework. "The neologism 'intra-action' signifies the mutual constitution of entangled agencies," she explains. "That is, in contrast to the usual 'interaction,' which assumes that there are separate individual agencies that precede their interaction, the notion of intra-action recognizes that distinct agencies do not precede, but rather emerge through, their intra-action."[37] Barad's theorization of intra-action reworks traditional ontological notions of causality and challenges binary relationships. This calls into question the preexistence and fixedness of any entity. Whether the phenomenon under investigation is an atom, a body, an experimental apparatus, a language, a scientific knower, or a collectivity, through their intra-action, each emerges in relation to other entities. Intra-action illuminates a relational ontology that is populated by mutually constituted phenomena.

Working with the categories of sex and gender in science, health, medicine, and engineering, feminist STS scholars do not ignore questions concerning the underrepresentation of women and other minorities in science. However, what has also become clear is that they have moved well beyond the "woman question in science." In 1985, Donna Haraway published "A Manifesto for Cyborgs: Science, Technology and Socialist Feminism in the 1980s," where she problematized essentialisms, including the idea of what it means to be a woman.[38] In 1986, Sandra Harding started moving us beyond the women in science question by challenging us to consider the science question in feminism.[39] As discussed in chapter 2, Sue Rosser followed up Harding's challenge by calling for more feminists to enter into the sciences and to discover what else we could know about the relationship between science and feminism.[40] As a feminist physicist, Barad responded to this call and used her knowledge and experience of thinking simultaneously with feminism and quantum physics to open up a fresh ontological terrain that troubles the ontological distinctions between sex and gender.

Conjugating Biology and Culture

Feminist STS scholars have identified a host of theoretical and empirical problems that emerge when sex and gender are placed in a binary. For some time now this scholarship has called attention to such unsavory consequences of the uncritical deployment of the sex/gender binary, while also questioning the viability of the biology/culture distinction.[41] Drawing on ontological lessons brought forward by poststructuralist debates, as well as developments in biology and physics, these scholars have argued that biology and culture cannot be easily separated. Laying out a rich genealogy of the role and contribution of intersexuality in our understandings of gender as a concept, David Rubin has summarized trajectories of intersecting work on intersex, biomedicine, and feminism. "Biological processes are not exterior to culture," he states. Rather than suggesting that the two simply interact, Rubin's survey on decades of feminist, queer, and intersex scholarship indicates that the category of "sex cannot be definitively disentangled from gender" (i.e., a Venn diagram might not work).[42]

This ambiguity between the categories of sex and gender, as well as biology and culture, has been quite troublesome but also productive for feminist scholarship. In *Gender Trouble*, Judith Butler posed a series of far-reaching questions:

> Can we refer to a "given" sex or a "given" gender without first inquiring into how sex and/or gender is given, through what means? And what is "sex" anyway? Is it natural, anatomical, chromosomal, or hormonal, and how is a feminist critic to assess the scientific discourses which purport

> to establish such "facts" for us? . . . Are the ostensibly natural facts of sex discursively produced by various scientific discourses in the service of other political social interests? If the immutable character of sex is contested, perhaps this construct called "sex" is as culturally constructed as gender; indeed, perhaps it was always already gender, with the consequence that the distinction between sex and gender turns out to be no distinction at all.[43]

Putting aside for a moment the crucial ontological disruptions of sex and gender that Butler proposes here, she also raises the important question of how a feminist critic is to "assess the scientific discourses which purport to establish such 'facts' for us."[44]

Since Butler posed this question in 1990, the field of feminist STS has flourished. To be a feminist critic today does not automatically disqualify one from also being an informed scientist and/or feminist STS scholar who is able to knowledgably "assess the scientific discourses" and the facts that they "purport to establish." Much of this increased participation with the sciences is of course directly due to the efforts made by those molar projects in early feminist science studies that were dedicated to increasing the number of women and feminists in the sciences. Those efforts may have directly fed into the work produced by several feminists who went on to train in the sciences such as Barad. Interestingly, their grasp of scientific knowledge and intimate exposures to scientific practices have given rise to a subsequent generation of feminist scholars who can now attempt to answer Butler's questions regarding our treatment of sex. The next section highlights the work of two feminist scholars, Luciana Parisi and Myra Hird, who have done just this. Working closely with Lynn Margulis's work, both of these feminist scholars have used their exposure to bacteriology and the details of scientific research and practices to produce new insights for feminism. It is clear that feminists now have the tools to consider the question of whether sex is natural, anatomical, chromosomal, or hormonal, or whether the natural facts of sex are discursively produced. The answer of course is that it is all of these.

After more than three decades of critical engagement, feminist scientists and feminist STS scholars have learned a great deal about the scientific and biological treatments of the categories of sex and gender.[45] Many would agree that the use and understanding of the category "sex" in the discourses of the sciences remains complicated and highly nuanced, but that more often than not, sex (1) is treated as a given or as an ontologically fixed category; (2) is still either conflated with gender; and/or (3) is placed into a binary and interactionist relationship with gender (as seen in the *Gendered Innovations* project). This does not mean that feminist critics, who are qualified and able to rigorously assess the scientific discourses, wish to entirely dismiss the work of the biological sciences or sex differences research for that matter. For instance, Angie Willey has introduced the term "biopossibility" as "a species- and context-specific capacity to embody socially meaningful traits or desires," as well as "a tool for naturecultural thinking."[46] What scholars such as Willey, Rubin, and other feminist and queer STS scholars are interested in doing is challenging the supposed "immutable character of sex."[47] They do so not only by showing how cultural ideals and language play a role in shaping our understanding of biology and matter, but also by recognizing the contributions that biological matters make in shaping our cultural expressions and grasp of what we come to know as sex.

Thinking with biophilosophies of becoming, we can begin to see that biology and biological processes need not be essentializing for feminists. Importantly, however, such ontological and ethical gestures should not distract us from also examining the "political and social interests" that shape the sciences and our scientific discourses of sex.[48] As Rubin has stated, we understand that sex and gender are both associated with a biological materiality and cannot be definitively disentangled from each other or from culture. The question now is, How do we begin to reframe sex?

A Thousand Tiny Bacterial Sexes

In *A Thousand Plateaus*, Deleuze and Guattari discuss the idea of segmentarity, proposing that it exists principally in two modes—namely, a rigid mode and a supple mode. The rigid mode reflects an arborized state and molar view of segmentarity, whereas the supple mode is rhizomatic, representing openness to the multiplicities of molecular processes. They argue that the "binarities" that form from within supple modes of segmentarity are the result of "multiplicities of *n* dimensions."[49] While describing rigid and supple segmentarities, Deleuze and Guattari state that it is incorrect to simply oppose these modes of segmentarity and that they too are entangled. I am most interested in how they develop their concept of segmentarity in relation to sex. They state:

> Every society, and every individual, are thus plied by both segmentarities simultaneously: one molar, the other *molecular*. If they are distinct, it is because they do not have the same terms or the same relations or the same nature or even the same type of multiplicity. If they are inseparable, it is because they coexist and cross over into each other. . . . In short, everything is political, but every politics is simultaneously a macropolitics and a micropolitics. . . . If we consider the great binary aggregates, such as the sexes or classes, it is evident that they also cross over into molecular assemblages of a different nature, and that there is a double reciprocal dependency between them. For the two sexes imply a multiplicity of molecular combinations bringing into play not only the man in the woman and the woman in the man, but the relation of each to the animal, the plant, etc.: a thousand tiny sexes.[50]

Elizabeth Grosz quotes these same lines from Deleuze and Guattari in her own theories on the body and in an effort to envision the deterritorialized body without organs (BwO). She is interested in using Deleuze and Guattari's approaches to becoming in order to develop a feminist project of rhizomatics. This project views subjectivity through the lens of molecular becomings and treats the body, above all else, as a corporeal event that is capable of transformation.[51] My intention is similar. By working with bacteria, I wish to better understand how the quality of changefulness is made possible by sex. More specifically, I am interested in what happens to our understandings of sex and gender when we allow them to cross over from molar treatments into the molecular. I am also interested in thinking with bacteria to better understand sex as an event that plays on and emerges from the "multiplicity of molecular combinations."[52]

Extending feminist theories of sex and gender in her book *Volatile Bodies*, Grosz notes that "feminists have exhibited a wide range of attitudes and reactions to conceptions of the body and attempts to position it at the center of political action and theoretical production." Differentiating "sexual difference" approaches from "egalitarian feminism" and "social constructionism," Grosz suggests that such thinkers as Luce Irigaray, Helene Cixous, Gayatri Spivak, and Judith Butler are particularly wary of the sex/gender distinction. Grosz argues that for these feminists the body is not accepted as a blank biological slate upon which culture or gender is projected. Rather, Grosz explains that feminisms of difference understand the body to be *active*. In contrast to universal notions of essences or categories, she suggests that the body is seen as a "cultural interweaving and production of nature." As such, there are irreducible differences not only between the sexes but also among members of the same sex. Turning to Deleuze to support her reconceptualization of the body, and to reframe the ontological distinctions that have been drawn between biology and culture, Grosz notes that a "Deleuzian framework de-massifies the entities that binary thought counterposes against each other: the subject, the social order, even the natural world are theorized in terms of the micro-processes, a myriad of intensities and flows, with unaligned or unalignable components, which refuse to conform to the requirements of order and organization. . . . Identities and stabilities are not fixed."[53]

The work of Luciana Parisi and Myra Hird offers examples of what we can come to know if we do learn to de-massify the binary that counterposes sex and gender. For some time now feminist STS scholars have questioned the stable ontological moorings associated with binary and categorical notions of sex and gender as well as biology and culture. They have started to elaborate on the idea of processual bodies and, as a result, are looking more closely at the materiality and material contributions made by biological matter. Using bacteria as their guide, both Parisi and Hird have emphasized the importance of understanding bodies and biologies as microprocesses, and sex in particular, as an event comprised of a multiplicity of differences.

In *Abstract Sex: Philosophy, Bio-Technology and the Mutations of Desire*, Parisi argues that in this age of information, cybernetics, and human reproductive technologies such as genetic engineering and cloning, we must reformulate our understandings of sex. In this new era of cybernetic capitalism, she claims, human sex has been separated from so-called "natural" reproduction, further blurring the distinctions we make between natural sex and artificial sex. This blurring further impacts the distinction between the biological and technological as well as between embodiment and disembodiment. Parisi proposes her idea of "abstract sex" as a micro-political framework that can address this blurring effect and that can work to integrate three main nonlinear layers within which sex is stratified—namely, the biodigital, the biocultural, and the biophysical. "Instead of re-articulating sex within a post-feminist critical framework where difference is no longer material," Parisi states, "abstract sex extends the feminist politics of desire by mapping the transversal mixing of information between bodies of all sorts (bacteria, vegetables, animals, humans and technical machines). Abstract sex proposes to tap into the kinetic ethology of tiny sexes that lay out a micropolitics of symbiotic relations between different levels of mutation of matter and desire."[54]

Parisi also turns to the work of Margulis, and explicitly to bacteria, emphasizing that sex is ancient and that it emerged as a way for "bacterial cells to repair their DNA damaged by intense solar radiations." As she points out, sex, as a mode of transmission and reproduction of information, is almost as old as the Earth itself. In what can also be interpreted as an unapologetically posthumanist move, Parisi makes it clear that as humans we are embarrassingly new to this Earth compared to bacteria and have very little appreciation of the multiplicities of sex. We live in the "Age of Bacteria," as Parisi emphasizes, and have much to learn from the sex lives and practices of bacteria.[55]

Drawing our attention to the multiplicities of molecular combinations that are present in bacteria, Parisi reminds us that "transgenic sex, the recombination of genetic material from two or more cellular bodies, constitutes the most ancient mode of genetic transfer on the biophysical stratum: bacterial sex."[56] So that we do not mistake this propensity for transfer, change, and multiplicity with life and proliferation alone and forget to associate genetic transfer with death and destruction also, Parisi reminds us of the full range of consequences of the microprocesses of sex by turning our attention to the relationship between viruses and bacteria. She states:

> Viruses attack bacteria and inject their genetic material into the cell. They hijack the bacterial genetic system and start viral replication. The bacterial cell eventually reaches a critical point, a threshold of change by breaking apart, and new viruses, produced by the combination of bacterial and viral genes, will spread into a new bacterium host. Such a virulent sex corresponds to bacterial abilities to trade genes by developing a

variety of metabolisms—including the use of metals—of which plants and animals have learned to use only few. Bacterial sex not only breeds new genes, but also manipulates the genetic composition of the bacterial body itself. . . . Transgenic sex puts up no resistance to mutations and affords no protection from contagion.[57]

The virology professor from my undergraduate days would be very pleased to see this no-nonsense treatment of virus-bacterial encounters, even though he would probably not be interested or impressed in the slightest by the link made above between viruses, bacteria, transgenic sex, molecular politics, and feminism. I, on the other hand, am absolutely thrilled by Parisi's success at bridging these fields of scholarship, by her smooth mixing of disciplinary vocabularies, and by her ability to steer us directly into the rich complexities and complications that come with thinking about sex otherwise. As should be clearly evident, in response to the query made by Butler in 1990, feminists can indeed trouble the ontological categories of sex and gender and also be informed feminist critics—not only of the humanities and social sciences but also of the biological, physical, and natural sciences.

In a similar show of interdisciplinary expertise, Hird has called for a "bacterial ontology" and "microontologies" to reinvent our understanding of sex and sexual difference.[58] In *Sex, Gender and Science*, Hird turns to bacteria to create a "nonhumanocentric position" and shows that by "paying attention to nonlinear biology it is possible to acknowledge that human bodies, like all living matter, physically actualize sex diversity."[59] As Hird states:

Our remote ancestors continue to promiscuously exchange genes without getting hung up on sexual reproduction. Bacteria are not picky, and will avidly exchange genes with just about any living organism anywhere in the world, including the human body. Thus bacteria are beyond the false male/female dichotomy of human discourse. Since bacteria recognize and avidly embrace diversity, they do not discriminate on the bases of "sex" differences at all. The bacteria that move freely into and within our bodies are already infinitely "sex" diverse. . . . So in the tired game of identity, I would choose neither goddess nor cyborg. I would rather be a bacterium.[60]

By suggesting that bacteria are infinitely sex diverse, Hird may be in favor of the statement made by *Gendered Innovations* that microbiology textbooks should not represent bacteria as being "sexed" organisms in a binary understanding of the term. After all, the naming of bacteria as male is based on the criteria that males are the "donor" cells that possess a fertility factor (F+) that allows for the extension of a sex pili, which in turn is required for conjugation. Alternatively, female bacterial cells are those recipient cells that lack the fertility factor (F-).

For some, the language and model used to describe bacterial conjugation in this way may resemble the development of male and female sexuality as articulated by psychoanalytic theory and Freud's explanation of the development of normal female sexuality through her lack of a penis and subsequent penis envy. However, Hird's work helps us to see that bacteria do in fact engage in sex, and that we could instead be using a bacterial ontology to reframe how it is that we think about sex. For instance, drawing from the work of Margulis, Deleuze and Guattari, Grosz, Parisi, Hird, and many others, we could be thinking about sex as a process or as an event through the framework of symbiogenesis, rather than as a stable factor that exists within a closed binary. Given the distance that feminist theory and interdisciplinary feminist STS work has come in reframing the sex and gender binary, it would be a shame not to incorporate this thinking into our textbooks. If we can learn to treat sex as a haecceity, as a multiplicity, or as an event, we may even be able to reconcile this fact with the idea that bacteria actually are "sexed" organisms—multiple times, perhaps even a thousand tiny times over.

In addition to drawing our attention to sex diversity in bacteria, Hird highlights the capacity that bacteria have to communicate with each other. In *The Origins of Sociable Life: Evolution after Science Studies*, Hird points to the growth patterns of bacterial colonies and to their collective ability to sense, process information, and regulate gene expression in each individual bacteria through various signaling mechanisms.[61] Describing the processes that bacteria use to communicate changes required for antibiotic resistance, she states, "In other words, bacteria make use of a collective epigenetic memory that can, for instance, track previous encounters with antibiotics: they collectively glean information from the environment, 'talk' with each other, distribute tasks, and convert their colonies into a massive "brain" that processes information, learns from past experiences, and, we suspect, creates new genes to better cope with novel challenges."[62] This ability to communicate indicates the complexity that is involved in bacterial signaling and it goes without saying that microbiologists and genetic engineers have developed a great interest and appreciation for this capability.

However, Hird also touches upon the difficulty that biologists often experience while trying to avoid anthropomorphizing these bacterial communicative processes.[63] She motions us toward Charles Sanders Peirce's original theory of semiotics and suggests that microbiologists would do well by trying to understand bacterial communication through biosemiotics. "Constituting a theoretical approach or frame," Hird explains, "biosemiotics 'is concerned with the sign aspects of the processes of life itself' gleaned through the relationships between sign, object and interpretant. According to biosemiotics, what organisms sense also has a meaning (food, predator, escape, sexual mate and so on). All organisms are born into a system of signs—a semiosphere—consisting of the totality of movements, odors, colors, chemical signals, touch and so on."[64] Hird raises a very crucial point here. Similar to Parisi, by suggesting that bacteria like all organisms are born into a system of signs, Hird is enacting a critical posthumanist move that works to decentralize the human. By turning to semiotics to explain the communicative behavior of bacteria, Hird is further engaging in a molecular politics that works to include bacteria and other nonhumans as users and participants in the practices of communication, language, and meaning-making.

What I find interesting here is that even while making this radical and immanent gesture, Hird herself is cautious about anthropomorphizing the communicative abilities of bacteria. What if while describing the collective epigenetic memory of bacteria, she did not place quotations around the word "talk" and instead merely stated that bacteria talk to each other?[65] If bacteria can indeed communicate, why do we feel the need to hesitate or somehow differentiate between what we do and what they do by saying that they "talk," but that we humans simply talk? In the same vein, can we say that bacteria write or do they only "write"? Questions such as these bring me to the last binary distinction that I wish to analyze in this chapter. What can we come to know differently about the relationship between the material and the semiotic, or between matter and language? The question is not whether bacteria communicate, but rather how we as humans learn to orient our encounters with bacteria when they communicate.

Conjugating Matter and Language

In *Meeting the Universe Halfway*, Barad lamented the following with regards to the relationship between matter and language:

> Language has been granted too much power. The linguistic turn, the semiotic turn, the interpretive turn, the cultural turn: it seems that at every turn lately every "thing"—even materiality—is turned into a matter of language or some other form of cultural representation. The ubiquitous puns on "matter" do not, alas, mark a rethinking of the key concepts (materiality and signification) and the relationship between them. Rather, they seem to be symptomatic of the extent to which matters of "fact" (so to speak) have been replaced with matters of signification (no scare quotes here). Language matters. Discourse matters. Culture matters. There is an important sense in which the only thing that doesn't seem to matter anymore is matter.[66]

I have often wondered about the audience that Barad was targeting when she wrote that statement. Likely, she was appealing to feminist theorists influenced by poststructuralism and cultural studies, or radical social constructivists who, at least since the days of the science wars if not before then, have been attempting to confront and unravel several binaries including that between matter and language.

I have also wondered if scientists working in a lab—a biology lab, for example—who may be ideal users for the *Gendered Innovations* website, would lament over the status of matter in the same way that Barad does. Would they say that matter doesn't matter any longer when they are setting up their experiments to work with animals, when they are killing animals, harvesting tissues, incubating cells, or isolating DNA and proteins? I don't think so. When the scale, space, and timeline of animals' sleep and wake cycles, hormone pulsatility, or peaks in protein synthesis dictate your daily activities as a scientist in a biology wet lab, it is "matter" and not language that seems to have been granted more power. In fact, this is precisely why in the *Gendered Innovations* epigraph, a reverse warning is put into place. Scientists who get their hands dirty with tissues, cells, molecules, and other physical properties of matter on a daily basis have to actually be reminded that language and representations matter. Indeed, some days as an interdisciplinary scholar with one foot in the humanities and the other in the biological sciences, the constant back and forth I experience while trying to deal with the binaries that are drawn between matter and language (or between matter and text) is enough to give me whiplash.

In their book *New Materialism: Interviews and Cartographies*, Rick Dolphijn and Iris van der Tuin describe new materialism as a new metaphysics, which reinterprets previous work and creates a "new tradition" that alters understandings of the past, present, and future. Characterizing new materialism as a "transversal" cultural theory, they suggest that it "does not privilege matter over text or culture over biology. It explores a monist perspective, devoid of the dualisms that have dominated the humanities (and sciences) until today, by giving special attention to matter, which has been so neglected by dualist thought." Although I would argue that matter has not been neglected in all forms of dualist thought, I am drawn to the ontological univocity that is implied with such a transversal theory. Claiming to provide an "immanent answer to transcendental humanism," new materialists are intent on disassembling powerful dualisms (including sex and gender) to "do justice to the 'material-semiotic' or 'material-discursive' character of all events."[67]

Multiple strands of feminist theory contribute to the new materialisms. Feminist philosopher van der Tuin situates new materialism in relation to older forms of feminist materialisms that championed monism and vitalism. Positioning new materialisms as "the inheritor of feminist standpoint theory," she traces continuities in feminist epistemological debates and philosophical engagements with materiality including historical materialism.[68] The term "new materialism" or "neo-materialism" has been credited to Rosi Braidotti, whose work on posthumanism and feminist theories of subjectivity provides a basis for a new materialist thinking.[69] New materialism also traces its origin to Haraway's conceptualization of the material-semiotic.[70]

Despite these long trajectories, it could be argued that recent scholarship under the name of feminist new materialism also grows out of cultural theory's more recent engagement with decades of work in feminist

STS. In fact, the material turn, as conceived in relation to poststructuralism and cultural theory, stems from a very specific conversation within feminist theory. The particular conversation I am referring to here is the one that supposedly privileged postmodern constructivism and the significance of language to such an extent that matter itself appeared to be discursively constituted. Judith Butler's work is often taken as emblematic of feminist theory's "flight from nature" or indeed its "failed materiality.[71] Criticizing the preoccupation with the discursive elements of bodies and power, material feminists as well as feminist new materialists have sought to mitigate the influence of poststructuralism's linguistic idealism on feminist theory and have turned in many cases to the "hard" sciences in an effort to get closer to matter.[72] In her efforts to illuminate the cultural and constructed aspects of "sex," Butler suggests that the body cannot be known outside of inscription or discourse. "To posit by way of language a materiality outside of language is still to posit that materiality," she famously wrote, "and the materiality so posited will retain that positing as its constitutive condition."[73] Critics have claimed that Butler's refusal of any distinction between sex and gender precludes the possibility of material expressions of and by the body.[74]

Correctly or incorrectly, many feminist new materialists have interpreted this emphasis on language, or on text, as an inherent inability to think about matter—whether that matter is coded as sex, biology, atoms, or nature. Although Butler's work has served as fertile ground for this criticism, several European feminist scholars have suggested that her work is emblematic of a larger problematic characteristic of a dominant strand of US feminist theory. Articulating a Eurocentric approach to new materialism, Braidotti, for example, has advanced a "friendly but firm criticism of American hegemony in feminist theory, . . . attempt[ing] to develop other perspectives, drawn from historical and situated European traditions."[75] Braidotti and other European new materialists point specifically to misleading interpretations of Simone de Beauvoir's work on sex, gender, and sexual difference spurred by Butler's article "Sex and Gender in Simone de Beauvoir's Second Sex."[76] In contrast to Beauvoir's complex account, critics have charged Butler with installing "a strict dualism" by overemphasizing the sex/gender split and attributing to Beauvoir "an oversimplified idea of language."[77] European new materialists seek to rescue Beauvoir's work by foregrounding the undecidability of sexual difference in her works and by promoting her ideas of "sexual differing" and a "performative understanding of ontology."[78]

Vicky Kirby has also raised a number of critical challenges for feminist theory through her close readings of Butler's analyses of nature/culture, discursive ontology, and conceptions of materiality. Turning Butler's proposition that sex is "always already gender" on its head, Kirby asks the risky question, "What if culture was really nature all along?" She explores the possibility that signs are "substantively or ontologically material."[79] In a piece that can be found in several places, Kirby shares an exchange that she had with Butler that pushes up against questions of language, matter, nature, and biology.[80] In her interview with Butler, Kirby states that she is interested in thinking about what it is that prevents us from considering "signs as substantively or ontologically material."[81] She asks Butler: "In the face of contemporary medical research on the body in genetics, the cognitive sciences (I'm thinking of the similarity between neural-net behavior and Saussurean linguistics), immunology, and so on, there is a serious suggestion that 'life itself' is creative encryption. Does your understanding of language and discourse extend to the workings of biological codes and their apparent intelligence?"[82]

Butler replies:

> I take it that [you] want to know from this question and the earlier one what my engagement with science is. And here the question seems to be: does my view of discourse include "biological codes." I confess to not knowing the literature to which [you] refer. [You] may need to take me through the theory that interests [you] here so that I might more intelligently respond. From my recent exposure to the work of Evelyn Fox-Keller, I would, however, say the following, reiterating what I take [your] view to be. There are models according to which we might try to understand biology, and models by which we might try to understand how genes function. And in some cases the models are taken to be inherent to the phenomena that is being explained. Thus, Fox-Keller has argued that certain computer models used to explain gene sequencing in the fruit fly have recently come to be accepted as intrinsic to the gene itself. I worry that a notion like "biological code," on the face of it, runs the risk of that sort of conflation. I am sure that encryption can be used

> as a metaphor or model by which to understand biological processes, especially cell reproduction, but do we then make the move to render what is useful as an explanatory model into the ontology of biology itself? This worries me, especially when it is mechanistic models which lay discursive claims on biological life. What of life exceeds the model? When does the discourse claim to become the very life it purports to explain? I am not sure it is possible to say "life itself" is creative encryption unless we make the mistake of thinking that the model is the ontology of life. Indeed, we might need to think first about the relation of any definition of life to life itself, and whether it must, by virtue of its very task, fail.[83]

Kirby has followed up, in later reiterations of this exchange, by adding more background to the question she originally posed to Butler. Referring to the workings of "biological codes" and their intelligence, Kirby states: "On this last point, I was thinking of the code-cracking and encryption capacities of bacteria as they decipher the chemistry of antibiotic data and reinvent themselves accordingly. Aren't these language skills?"[84] In another reference to this interview, Kirby elaborates the point further by explaining:

> In a bid to illuminate why Butler's manoeuver will authorize the iteration of the problem she so carefully unpacks for us, namely, that ~~Nature~~ (now under erasure) is incapable of cognizing or reinventing itself, I asked her to consider a rather simple phenomenon. My question directly relates to the theme of this issues' problematic, namely, how to engage with science and its "objects." I was thinking about the cryptographic skills of bacteria as they decipher the chemistry of antibiotic data and reinvent themselves accordingly. When ciphering skills are exhibited by boffins such as Alan Turing of Enigma Code fame, Steve Wozniak, cofounder of Apple Computer, or the infamous "black hat" hacker in the nineties, Kevin Mitnick, we interpret this capacity for abstract thinking as an exemplary instance of intelligent reasoning. Although we are unlikely to describe the growing number of superbugs in terms of these same special talents, it could nevertheless be suggested that these single-celled microorganisms with no nucleus (or "head") have actually outsmarted their human interlocutors.[85]

No matter how many times I read it, I will be honest and admit that even with my undergraduate training in microbiology and graduate and postgraduate work in molecular biology, I struggle to grasp what Kirby means by the ability of bacteria to "decipher the chemistry of antibiotic data." I consider myself somewhat fluent in the characteristics and functions of bacteria that can be found in any standard microbiology or bacteriology textbook. I am also familiar with the processes by which antibiotics target peptidoglycan synthesis to break down bacterial cell walls.[86] I am aware that bacteria have several mechanisms for developing resistance to antibiotics including the modification of protein structures that interfere with antibiotic carriage into the cell, that there are changes that occur in bacterial genomes that can cause antibiotic resistance by spontaneous mutation, and that these beneficial mutations can be passed through both horizontal and vertical genetic exchange.[87] Yet, without a single reference to the vast scientific literature on bacterial resistance to antibiotics, what Kirby means by "the chemistry of antibiotic data" (which was meant to further clarify her original question to Butler) remains unclear to me.

Despite this lack of scientific clarity, I am convinced that Kirby is doing vital work by suggesting that life itself is creative encryption and by calling our attention to the capabilities of bacteria and the intelligence behind their biological codes.[88] By asking whether the skills that bacteria have are language skills, Kirby is attempting to use scientific knowledge to contribute to the evolution of key concepts in feminist theory that are grounded in the sex/gender binary. She is trying to force a particular audience of feminist theorists who have been trapped in a transcendent humanistic frame, and who have moved away from directly addressing questions of biological and physical matters, into rethinking their approach to matter. Following Jacques Derrida, Kirby forces the question of whether a distinction can be made at all between matter and language, matter and text, and matter and writing (writing understood here in a broad sense).

Interestingly, in order to support this posthumanist effort to acknowledge the communicative capacities of nonhumans, and to consider life itself as a text, Kirby compares two contrasting interpretations of Derrida's claim that "there is no outside of text."[89] The first interpretation reflects the view of many critics of poststructuralism who assert that "we are caught in an endless slide of referral that leads from one signifier to another signifier, one meaning to yet another meaning, in a vertiginous spiral of implication that never quite arrives at its destination. As a consequence, we can never retreat or advance to some natural, prediscursive, or extratextual space in order to test the truth or adequacy of our representations because, as we have seen, intelligibility is reckoned through such systems."[90]

As an alternative interpretation, Kirby appeals to Derrida's claim "there is no outside of text" rather as an attempt to grasp "the worlding of the world" by thinking about writing in a more general sense. She argues that "'writing in the general sense' articulates a *differential* of space/time, an inseparability between representation and substance that rewrites causality. It is as if the very tissue of substance, the ground of Being, is this mutable intertext—a 'writing' that both circumscribes and exceeds the conventional divisions of nature and culture. If we translate this into what is normally regarded as the matter of the body, then, following Derrida, 'the most elementary processes within the living cell' are also a 'writing' and one whose 'system' is never closed."[91] Thus Kirby makes a compelling case that cells write and that "it is in 'the nature of Nature' to write, to read and to model."[92] The ontological openness of this stance, treating writing as a process that exceeds the divisions of nature and culture, accommodates feminist theories such as Haraway's conception of naturecultures and Braidotti's accounts of posthumanism that have also had formative impacts on the field of feminist STS.[93] In my estimation, thinking about the communicative capacities of bacteria via their writing skills has several hallmarks of being a project for molecular feminisms.

Thanks to my years in the lab and to experiences of "engaging with science and its 'objects'" such as neurons and microorganisms including bacteria, I am also absolutely on board with the idea that bacteria have special talents, including the ability to write, and not just "write." I am persuaded not only by Kirby's posthumanist gesture to extend capacities for language to nonhumans but also by her idea of seeing language, writing, and text itself as life. In fact, in chapter 5, I go one step further and extend these capacities to a concept of life that is not reduced to the organism or to the organic.[94] I am more than willing to go here with Kirby. However, I suggest that if we want to consider her claim of bacterial writing through biophilosophies of becoming and microphysiologies of desire, we must begin to connect such an ontological gesture to an applied ethics of matter.

The Microprocesses of Bacterial Ethics

As Nancy Tuana has stated, "It is easier to posit an ontology than to practice it."[95] Saying that bacteria write is important, but our work does not end here. The challenge lies in learning how to identify and appreciate the material consequences that accompany such an ontological position. At its best, feminist STS is immersed in analyzing the practices of and data generated by specific sciences, along a wide array of different scales. In fact, the richness and credibility of STS itself depends on developing systematic knowledge of the minute details of specific fields in the sciences while also keeping an eye on the larger organizational, institutional, and political structures associated with the circulation of power. Many of the scholars who engage more closely with the sciences are keenly aware that there are material consequences of our particular ontological conceptions of the material world, which have profound implications for species other than our own. As Stacy Alaimo and Susan Hekman poignantly state in their introduction to *Material Feminisms*:

> Redefining the human and nonhuman has ethical implications: discourses have material consequences that require ethical responses. Ethics must be centered not only on those discourses but on the material consequences as well. . . . A material ethics entails . . . that we can compare the very real material consequences of ethical positions and draw conclusions from those comparisons. We can, for example, argue that the material consequences of one ethics is more conducive to human and nonhuman flourishing than that of another. Furthermore, material ethics allows us to shift the focus from ethical principles to ethical practices. Practices are, by nature, embodied, situated actions. Ethical practices, which unfold in time and take place in particular contexts, invite the recognition of and response to expected as well as unexpected material phenomenon. Particular ethical practices, situated both temporally and physically, may also allow for an openness to the needs, the significance, and the liveliness of the more-than-human world.[96]

With this emphasis on considering the needs and liveliness of the more-than-human world, let us return to the ontological status of bacteria that write and to possible microphysiologies of desire that can serve as an applied ethics of matter as we go forward. What happens when a notion of language is extended to Nature? How do we begin to encounter this Nature? How do we encounter bacteria that we now acknowledge as having the capability to write? There are good reasons for recognizing that bacteria have special talents, of which writing may be one. However, as Alaimo and Hekman suggest, the statement Kirby makes, that "it is in 'the nature of Nature' to write, to read and to model," has ethical implications, precisely because it redefines the human as well as the nonhuman.[97] If it is accepted that Nature writes, reads, and models and that signs are "substantively or ontologically material," we should also be equally concerned with the outcomes or consequences of this ontology.[98]

It is one thing to say that bacteria write. It is another to learn how to pay attention to how bacteria write, why they write, and what they are writing. In fact, although Butler's response to Kirby's initial question regarding "life itself" as encryption and the "workings of biological codes" reveals Butler's belief in a materiality but our ultimate inaccessibility to this materiality through language, we must also point out that Kirby does little to address the questions about context that Butler posed back to her during their exchange. It is this emphasis on looking for both specificity as well as keeping an eye on broader contexts that I wish to extend to Kirby's references to the "code-cracking," "encryption capacities," and "cryptographic skills" of bacteria. Perhaps in a way it is an attempt to address Butler's poignant question that she posed to Kirby: namely, "What of life exceeds the model?"[99]

If we turn toward bacteria that write by *feeling around for the organism*, we begin to explore what a bacterial ethics might look like. We might ask, for example, how are bacteria changed, how are labs changed, how are institutions changed, and how are we as human scientists changed when bacteria begin to write? We might ask, for example, if genetic engineers, who think that do-it-yourself (DIY) synthetic biology represents a democratic science, are willing to extend coauthorship and co-ownership not only to other human scientists but also to the bacteria that perform most of the writing and biolabor in synthetic biology. We can begin to see that the model that presents DNA as code has already become the ontology of life. A molecular line of questioning would have us ask whether we are willing to work within this ruin. Are we willing to work with bacterial DNA as code to better understand how bacteria alter their immune systems in order to help humans fight their own diseases through genome editing? Biologists, geneticists, and bioengineers have long afforded bacteria with the capabilities of writing. They are beyond asking the question of what bacterial writing *is* and are more attuned to the question of what bacterial writing *can do*. In fact, they have built entire disciplines and biotechnologies based on the ontological premise that life, in the form of DNA, is text and that organisms such as bacteria not only write; they also transcribe, translate, and edit. Indeed, the model has long been accepted as the ontology, and although I am aware that this ontology needs to be constantly interrogated, I am also aware that biotechnologies that are based on the writing and editing capabilities of bacteria already exist.

By following our inquiries and engagements with bacteria through microphysiologies of desire, we may begin to ask different kinds of questions regarding these intimate encounters with bacteria. We could begin to ask whether in the specific case of bacteria, writing, reading, and modeling "allow for an openness to the needs, the significance, and the liveliness of the more-than-human world," or whether our acknowledgment of bacterial writing works to support, promote, and benefit only the most humanist of causes?[100] Are we at all concerned with the different genres of bacterial writing that might exist? If we can think about bacterial writing, can we also start to think about bacterial poetry? Or is bacterial writing only valued for its mechanistic appeal?[101] Can we only learn to recognize or appreciate bacterial writing when it serves a mechanistic function such as building antibiotic resistance and interrupting genes that contribute to human diseases, or when it is aligned with militarized or highly gendered skills such as cryptography, hacking, and code-cracking? Does this new ontology ultimately serve as a "reshaping" for "productionist purposes?"[102]

Trying to address the inseparability between representation and substance, Kirby suggests writing as a mutable "intertext." Although I read this mutability as a feature resonating with the molecular capacities for changefulness and nonhuman becomings, I want to keep in mind that biophilosophies of becoming prompt us to examine the specific "contexts" in which such moments of "intertext" occur. To gain more insights into the importance of thinking with bacteria and to better contextualize the worlding of the world through bacterial writing, it is perhaps helpful to return one more time to Derrida. It could be argued that Derrida's phrase written in French, "*il n'y a pas de hors-texte*," is better translated in English as "there is no outside-text," rather than "there is nothing outside of the text," or Kirby's use of the phrase "there is no outside of text."[103] In his own later work *Limited Inc*, Derrida noted that "the phrase which for some has become a sort of slogan, in general so badly understood, of deconstruction ('there is nothing outside the text' [*il n'y a pas de hors-texte*]), means nothing else: 'there is nothing outside context' or that 'nothing *exists* outside context.'"[104] It could be that Kirby's use of a particular translation of Derrida's phrase has influenced her interpretation of Derrida and where she chooses to turn her thoughts in relation to bacteria, bacterial writing, and the "worlding of the world."

Kirby's ontological intervention in extending language, reading, and writing skills to the nonhuman is crucial. However, in the worlding of the world that biophilosophies of becoming would have us bring forward, context matters. We must realize though that just as one cannot fix the meaning of a text, of course what counts as context also cannot be fixed. It depends on how an event is oriented. The final section of this chapter treats bacterial writing as an event, one that can be reframed and understood through postcolonial and decolonial haecceities. It interrogates the implications that follow when we say that bacteria have the capabilities to write. As Parisi reminds us, we must keep in view the full range of consequences that accompany the microprocesses of becoming, whether we are talking about bacterial sex or bacterial writing. As crucial and productive new ontological approaches evolving out of the interrogations of sex/gender, biology/culture, and matter/language binaries can be, we must remember that they don't exist outside context.

Bacterial Writing as an Event

Postcolonial and decolonial STS work to decolonize relations and practices. While raising the question of whose interests are served by venturing toward new ontological terrains, they emphasize the importance of giving voice to a broader range of knowledge bases as we produce these accounts. Despite all the recent attention to the so-called "material turn," many postcolonial and decolonial scholars recognize that certain bodies—such as people of color, reproductive bodies, disabled bodies, animals, plants and others subjected to colonialism, racism, capitalism, patriarchy, and science—have been inextricably tied to "nature" and never had the opportunity or the interest to join the so-called feminist "flight from nature."[105] Postcolonial and decolonial STS emphasize that all technologies should be contextualized. In fact, every technology, including bacterial writing, can be thought of as an event that is connected to specific traditions and practices of knowledge production. To decolonize these relations and practices, we can begin by reframing bacterial writing as an event. This event can be placed within postcolonial and decolonial haecceities, which encourage us to consider all technological events in relation to (1) transnational processes of colonialism and imperialism; (2) capitalist practices of production, consumption, and commodification; (3) gendered and raced labor of production and reproduction and the abstraction of this labor; (4) neoliberal forms of individualism and imperialism; and (5) effects of technology on global as well as local scales.

Recent work on the role of bacteria in the new life sciences demonstrates why thinking about bacterial writing as an event is crucial. For instance, in *Biocapital: The Constitution of Postgenomic Life*, Kaushik Sunder Rajan analyzes two important domains of current global consciousness. The first includes the life sciences, which he argues are "increasingly becoming information sciences." The second is capitalism, which he suggests has "defeated alternative economic formations such as socialism or communism and is therefore considered to be the 'natural' political economic formation, not just of our time but of all times." As an STS scholar, Sunder Rajan insists that the life sciences and capitalism are coproduced, stressing that the "life sciences are *overdetermined* by the capitalist political economic structures within which they emerge." He explains: "'Overdetermination' is a term used by Louis Althusser to suggest a *contextual* relationship, but not a *causal* one (Althusser 1969 [1965]). In other words, even if a particular set of political economic formations do not in any direct and simplistic way lead to particular epistemic emergences, they could still disproportionately set the stage within which the latter take shape in particular ways."[106] Sunder Rajan's argument can be read to suggest that to discuss bacteria as writers, readers, and modelers, is to also discuss them in a contextual relationship—one that is structured by the domains of the life sciences and capitalism. Similar to their human, animal, and plant counterparts, in the purview of life sciences, bacteria not only carry information, they *are* information. This particular ontological formation has influenced epistemic emergences for bacteria. For bacteria the sign does become ontologically material, but as humans it appears that we are only willing to recognize this through bacterial labor.

However attractive it might be to imagine the bacterial capabilities of writing, in the life sciences that are driven by contemporary forms of capitalism, these capabilities are perceived through a highly mechanistic view of life. A view that not just figuratively but literally forces bacteria to write. In the realm of synthetic biology, for example, bacteria such as *Escherichia coli* are being purposely bred for their writing capabilities. This writing consists of transcribing DNA and translating RNA for human interests alone. In one case, students participating in the annual iGEM (the International Genetically Engineered Machine) competition in synthetic biology have designed *E. coli* to produce a wintergreen scent during their exponential growth phase, and a banana scent during their stationary phase to address the olfactory challenge that scientists in molecular biology labs often have to face while working with bacteria. To be clear, it is precisely because of humans, who work in molecular biology labs with bacteria and do not want to be offended by their odor, that bacteria have been forced to rewrite their own genomes. As Sunder Rajan reminds us, even though the political economic formations that have led to genetic engineering technologies themselves may not have caused this event directly, we can begin to see the epistemic climate that these very formations produced, which have led to the design and emergence of banana-scented *E. coli*.

In another bacterial writing event, Melinda Cooper has suggested that the skills of lateral gene transfer through transformation, transduction, and conjugation have made bacteria of utmost interest to humans. This interest goes beyond studying bacteria for the human diseases they cause and for their antibiotic resistance. As Cooper notes, the observed ability of bacterial plasmid exchange and transformations led to the advent of recombinant DNA and genetic engineering technologies in the 1970s. Indeed, synthetic biology (discussed further in chapter 5) is but the latest face of genetic engineering. Instead of combining a single gene of interest into a bacterial plasmid, synthetic biologists have created entirely human-designed genomes of human interest that they have then inserted into "surrogate" bacterial cells. These bacterial cells, created explicitly for their writing capabilities are mass (re)produced in the lab. In her book *Life as*

Surplus: Biotechnology and Capitalism in the Neoliberal Era, Cooper poses the pivotal question, "Where does (re)production end and technical invention begin, when life is put to work at the microbiological or cellular level?"[107] She argues:

> Recombinant DNA (rDNA) differs from previous modes of biological production in a number of ways. First, while microbial biotechnologies such as fermentation are among the oldest recorded instances of biological production, recombinant DNA constitutes the first attempt to mobilize the specific reproductive processes of bacteria as a way of generating new life forms. Moreover, recombinant DNA differs from the industrial mode of plant and animal production in the sense that it mobilizes the transversal processes of bacterial recombination rather than the vertical transmission of genetic information. This is a technique that lends itself to the specific demands of post-Fordian production—flexibility and speed of change—to a degree that was impossible in traditional plant breeding.[108]

Discussing "biological growth" in the context of neoliberal biopolitics, Cooper argues that "neoliberalism reworks the value of life" by "effac(ing) the boundaries between the spheres of production and reproduction, labor and life, the market and living tissues."[109]

In the event that is bacterial writing, it is impossible to isolate and appreciate this skill outside of the political and economic formations of production, reproduction, and labor. It is important to keep in mind that in most cases, once bacteria have performed their tasks of transcription, recombination, or gene editing, they are promptly snuffed out for the valuable proteins they carry inside of them. This is not the high-profile secret life of cryptographers and code-breakers that we would like to imagine. More akin to the value that was assigned to "natives in the jungle" during colonial science, bacterial writing becomes bacterial labor, which is "harnessed and controlled," and "its products managed and turned into profitable use through the imposition of order and predictability."[110]

New ideas, politics, and practices can emerge at the intersections of molecular biology and feminism when we attempt to think with bacteria. By reflecting on what bacteria are capable of doing, we begin to change our understanding of some dominant binary relationships such as sex/

gender, biology/culture, and matter/language. We begin to see that bacterial desires, responses, experimentation, and communication skills such as writing can change how we frame our ontological, epistemological, and ethical questions in the lab. We begin to see the importance of thinking about our technological futures through bacterial lives. We may never come to know bacteria completely, but at least we can try to reframe our encounters with them.

4

Should Feminists Clone? And If So, How?

> For these are strange times, and strange things are happening.
>
> —ROSI BRAIDOTTI

> The idea is to build our own transporting machine and use it to get a relay going and to keep it going, creating ever greater and more powerful amalgamations and spreading them like a contagion until they infect every identity across the land and the point is reached where a now all-invasive positive simulation can turn back against the grid of resemblance and replication and overturn it for a new earth.
>
> —BRIAN MASSUMI

Several years back, I seriously thought about cloning myself, but only twice. I needed two duplicates, plus an original template, making for a total of three. The plan was that one clone could teach in a women's studies department and another clone could raise two young children. The third, the original me (the "template"), could take care of such pleasantries as writing this book. I, the template, would also be responsible for developing feminist STS practices so that the pleasures and dangers of new and emerging biotechnologies such as cloning would not go unexamined, as Rosi Braidotti put it, in these "strange times."

As a feminist scientist, I have been interested in developing projects in feminist STS through molecular politics. In their theory of becoming, Gilles Deleuze and Félix Guattari suggest that "all becomings are already molecular" and that on our way to becoming imperceptible, we must "always look for the molecular, or even submolecular, particle with which we are allied."[1] However, this interest is not altogether detached from mainstream molar projects that may be more recognizable to women's movements or other identity-based social justice work. In fact, I argue that in many cases, without molecular projects that emphasize biophilosophies of becoming and advance an immanent ethics of matter, our biological bodies, organs, cells, and molecules can be made to work against the very molar projects that rely on stable categories such as that of "Women." As Deleuze and Guattari have stated, "it is thus necessary to conceive of a molecular women's politics that slips into molar confrontations, and passes under or through them."[2] This being said, molecular positions are often risky and can be potentially provocative, with the processes of becoming, from woman to animal to molecules, appearing foreign (if apparent at all) to more conventional molar projects or what Isabelle Stengers has referred to as "moral" projects.[3]

I have posed the question, and quite boldly I might add, Should feminists clone?[4] The first time I asked this question, I was sitting on a panel surrounded by other feminist scholars.[5] The looks of horror following my query told me that I had hit a nerve. Hadn't I read Gena Corea's *The Mother Machine*?[6] Wasn't I aware that women's bodies have historically been used to support new reproductive and genetic technologies? Didn't I realize that cloning was the latest iteration of a history of technologies and scientific knowledge-making traditions that worked to oppress women through their biology? What became obvious to me was that although this history and context was absolutely important to remember, as the only trained biologist on the panel, I was interested in pursuing a different kind of politics.

The question had come to me during a time in my PhD research when I was conducting experiments on an *in vitro* cell line of hypothalamic neurons, investigating the possibility of feedback regulation of these neurons by gonadal steroids and by the pineal hormone melatonin. My experiments involved using a molecular biology–based technology referred to as *subcloning*. I worried that using subcloning in my research, a technology that feminists on the panel were obviously opposed to, would disqualify me from being a feminist. Now, several years later, the question not only lingers in my mind but has in fact grown into a monster of sorts, feeding off technological, organic, and political fears and hopes.[7] Even back then, however, I knew that in some ways my desires to participate in reproductive biology research and to use molecular biology–based technologies were somewhat "inappropriate" to some feminist ethical orientations. The ethical orientations I refer to here generally follow feminisms of equality that would rather focus on molar issues of being and identity and work toward treating women as liberal human subjects. Feminisms of equality aim to correct the conditions whereby women are considered less than human. Understandably, this ethical position—which strives to safeguard women, protect their reproductive rights, and fight for their equal rights—can be opposed to a great deal of molecular and reproductive biology research and technologies.

I believed then, and still do, that to deal with our posthuman living conditions, some feminists must express their energies and desires for change in different ways. Not everyone should or needs to work toward achieving subjectivity in that liberal humanistic sense. Instead, some of us must also turn our attention to developing molecular projects. Explaining what kinds of ethics are possible for postmodern subjectivities, Braidotti states:

> Ethics in poststructuralist philosophy is not confined to the realm of rights, distributive justice, or the law, but it rather bears close links with the notion of political agency and the management of power and of power-relations. Issues of responsibility are dealt with in terms of alterity or the relationship to others. This implies accountability, situatedness and cartographic accuracy. A poststructuralist position, therefore, far from thinking that a liberal individual definition of the subject is the necessary precondition for ethics, argues that liberalism at present hinders the development of new modes of ethical behaviour.[8]

I knew that by participating in the production of scientific knowledge on the body, I could use my micropolitics to new ends. I wanted to produce knowledge that addressed my concerns around contraceptives, hormone replacement therapies, and new reproductive and genetic technologies. This is why I decided to pursue a PhD in reproductive biology, why I worked with hypothalamic neurons of the brain in Petri dishes, and why I searched for the presence of estrogen receptor proteins using the technique of subcloning.

By thinking with molecular feminisms, molecular biology, and the question of cloning, my intention has never been to leave aside women, women's health, or reproductive justice issues. Rather, I have wanted to bring them into closer zones of proximity with the sciences and develop new forms of kinships with other actants—human, nonhuman animal, multicellular, and unicellular organisms—all of whom have a share in this biotechnological future. As a biotechnology, cloning is borne out of research in molecular, developmental, and reproductive biology, disciplines that have tried very hard to discipline biologies, bodies, and molecules. Dysfunctional as it may be, the culture of cloning has formed a new set of kinship arrangements, one that resembles a mess of growing crabgrass more than a neat and linear family tree and brings together disparate bodies starting from bacteria to plasmids, genes to eggs, uteruses to fetuses, humans to machines, and whole bodies to supposedly more identical whole bodies. Thinking through biophilosophies of becoming, this chapter considers the qualities of kinship and hylozoism more closely. It attempts to use these qualities to develop a strategy for feminist scientists to navigate their way through this strange era of cloning. It attempts to use the biotechnology of cloning to overturn itself, if not for a new earth, as Brian Massumi suggests is possible, then at least for a fresh set of politics.

What politics can emerge when we pay closer attention to the lab protocols involved in cloning? What happens when we learn to make kin with the nonhuman knowers and doers who actually carry out these biological processes for us in the lab? Building on the previous chapter, the skills of bacteria and bacterial plasmids are placed on an equal footing with human scientists, who have only recently learned how to recombine DNA in a lab. I apply a hylozoic view to the microorganisms and molecules that make molecular biology research and molecular biotechnologies possible in the first place. More important, I examine what happens when as scientists we become open to the push and pull exerted on us by our microscopic kin.

Tactical Recombinant Technologies

In his essay "Realer Than Real: The Simulacrum According to Deleuze and Guattari," Massumi tells us that according to Jean Baudrillard, "we breathe an ether of floating images that no longer bear a relation to any reality whatsoever" and that this is "simulation: the substitution of signs of the real for the real."[9] According to Massumi, Baudrillard can be interpreted to be lamenting the loss of the real and figures the simulacrum as a copy of a copy without any reference to an external model. Massumi suggests that such sentiments can make sense if the only "alternative to representative order is absolute indetermination."[10] Massumi states:

> Baudrillard's framework can only be the result of a nostalgia for the old reality so intense that it has difformed his vision of everything outside it. He cannot clearly see that all the things he says have crumbled were simulacra all along. . . . He cannot see becoming, of either variety. He cannot see that the simulacrum envelops a proliferating play of differences and galactic distances. What Deleuze and Guattari offer, particularly in *A Thousand Plateaus*, is a logic capable of grasping Baudrillard's failing world of representation as an effective illusion the demise of which opens a glimmer of possibility. Against cynicism, a thin but fabulous hope—of ourselves becoming realer than real in a monstrous contagion of our own making.[11]

If one accepts the philosophical system of positivism and the possibility of achieving aperspectival objectivity, it follows that what biologists do while working in a lab somehow gives us access to a "real nature." This is perhaps why with the advent of molecular recombination and cloning techniques, many are lamenting the loss of this old reality and our grasp of "original nature" through biology. With recent developments in the field of synthetic biology, however, scientists have long moved away from efforts of preserving an "original model" to understand the molecular basis of "naturally occurring" life and have instead wholeheartedly embraced the idea of the simulacrum.

These scientists are more interested in what *biology can do*, regardless of being an original or a copy. They are not held back by the virtues of the authentic or the significance of a "real" in biology. Since the discovery of the structure of DNA by James Watson and Francis Crick in 1953 (made possible by Rosalind Franklin's work), scientists have witnessed the process of DNA replication and the endless winding and unwinding of the double helix. Perhaps for this reason they are perfectly at ease in grounding their scholarly and industry-driven biotechnological pursuits in a world that deals entirely in the realm of the simulacrum, in a world inundated by copies of copies. DNA replication—this is what biology can do. Therefore, what does the significance of a copy of a copy matter when, for instance, a replicated DNA transcript of a previously transcribed and sequenced PCR product can be inserted into a plasmid vector and used to make an organism fluoresce green? Molecular biology, knowingly or not, is already deeply embedded in the hyperreality of simulation. It has moved beyond the "question of substituting signs of the real for the real."[12]

Given my interest in molecular feminisms and the question I posed regarding cloning practices, I am curious as to how, in this world of copies of copies, Massumi suggests we can we look for that "glimmer of possibility." "The challenge is to assume this new world of simulation and take it one step farther," he states, "to the point of no return, to raise it to a positive simulation of the highest degree by marshaling all our power of the false toward shattering the grid of representation once and for all."[13] In an effort to take on this challenge, this chapter makes the risky argument that recombinant DNA technologies such as subcloning may be used to dismantle the grid of representation and move us toward a playful proliferation of differences. Admittedly, I am drawn to thinking more closely about the radical potential of recombinant DNA technologies as tactical tools for feminist scientists, due to my own fond associations with bacteria, plasmids, and subcloning in the lab.

For me, the most pleasurable part of creating recombinant DNA was the polymerase chain reaction (PCR) procedure. I would begin by pipetting 5 microliters of mineral oil on top of the reaction mixture that contained the oligonucleotides, DNA and Taq polymerase, to avoid evaporation. The PCR machine would be warm to the touch, and the mineral oil flowed lightly and smoothly out of the micropipette tip into the tiny Eppendorf tubes (epis). Sometimes, I had a chance to add tiny drops of mineral oil into each crevice of the thermal cycler temperature block, watching the oil melt down into an even thinner form. From the warmth of my nonlatex gloved hands to the warmth of the thermal block, I imagined that each epi would be gently gripped by the mineral oil in the holes of the thermal block. I would bide my time through the steps of denaturation, annealing, and extension, knowing that the warmth would end abruptly once the PCR machine entered the cooling cycle and performed its final hold. Once the reactions had taken place and the temperature of the thermal block had dropped to somewhere between 4°C and 15°C, I would remove the epis onto a plastic rack that was destined for the –20°C freezer. Although the contents within the Eppendorf tubes looked completely identical to how they looked before they were placed into the PCR machine, I knew that the DNA and enzymes, although thought of as "raw biological materials," had done their work. This idea that matter, as "raw" as it may be, is also capable of some form of expression or movement, defines the quality of hylozoism. In addition, similar to the production of simulacra, it was through these multiple processes of replication and repetition that the potential existed for something in the DNA to change, thereby introducing a "proliferating play of differences."[14]

I could tell that in the processes of cloning and repetition, there was not simply the lost air of simulation but also perhaps the opportunity for emerging breaths of difference. With this knowledge I went about amplifying DNA on a regular basis for years, wondering how to move from repetition to positive repetition, or the inhabitation of a dominant discourse in order to open up a new site. Recombinant DNA therefore might be thought of as a simulacrum that has successfully broken out of the copy mold. Made through a series of repetitions and slight modifications, it has acquired a new purpose and function. Like the simulacrum, recombinant DNA "is less a copy twice removed than a phenomenon of a different nature altogether" undermining "the very distinction between copy and model."[15] In a Deleuzian ontology, whereby repetition is associated with difference, recombinant DNA technologies may allow for the emergence of "new experiences, affects and expressions to emerge."[16]

Tropes and Turns

For some time now, in some areas of feminist theory, there has been a call for feminists to find their way back to the matters of the body and back to biology. Elizabeth Wilson, for instance, posed the following questions: "How many feminist accounts of 'the anorexic body' pay serious attention to the biological functions of the stomach, the mouth or the digestive system? How many feminist analyses of 'the anxious body' are informed and illuminated by neurological data? How many feminist discussions of 'the sexual body' have been articulated through biochemistry?"[17] Many feminists have answered this call by turning to the practice of re-reading earlier scientific works, such as those belonging to Darwin and Freud, in order to diffract important biological theories for feminist purposes.

Although these re-readings and diffractions are important, a return to biology must involve more than the generous exploration of already established biological theories and their relevance to feminism. The "project of refiguring the biological" must also support the movement of feminists back into the laboratory for the production of *new* biological theories.[18] In order to be "informed and illuminated" by current neurological data or biochemistry in more than a cursory manner, feminists also have to actually learn the science and the scientific techniques, then attempt to make new scientific knowledges.[19] I have argued that to do this, feminists must learn how to face the nitty-gritty technical core of scientific knowledge production as well as all the contradictions, tensions, and dilemmas that come with carrying out scientific experiments. Part of the conversation that has been missing from this feminist call to return to the body and to biology, therefore, has to do with the lack of attention paid to developing feminist practices in the natural sciences. We need more ways to support feminists in the lab who can help us make this return to biology and give them the tools to deal with the dilemmas they will face upon this return. This is precisely why I posed my risky question regarding subcloning in the first place and why I return to it now. Despite its inappropriateness, I am interested in exploring the possibility that the scientific practices involved in the processes of cloning may be used to create new feminist politics.

Indeed, the idea of cloning can produce several discomforts for feminists. Some of this discomfort may be attributed to the fact that "cloning" serves as a particularly popular trope and is prone to metaphorical use. The use of metaphors, of course, holds an important place in scientific endeavors. As Donna Haraway has said: "A metaphor is the vital spirit of a paradigm (or perhaps its basic organizing relation). . . . It leads to a searching for the *limits* of the metaphoric system and thus generates the anomalies important in paradigm change."[20] I first posed my cloning question with the intention of finding out whether feminist scientists such as myself should conduct molecular biology experiments using the technique referred to as subcloning. As a biotechnology, subcloning has been used in the processes of directed evolution and has been integral to molecular biology research. The suggestion that metaphors lead to paradigm change is absolutely crucial to my project. Regarding the use of metaphors, Joseph Rouse has stated: "When tropes work, they 'turn' us, cause us to attend to or respond to things differently. . . . Tropes stand out from other unexpected uses of words in the responses they evoke: they amuse, provoke, associate, resound, and so forth . . . but they can also be occasions for reconfiguring the connections among things, utterances, and other practices."[21]

My intention in this chapter is twofold. First, I want to use cloning not simply as an entertaining metaphor but as part of a molecular project that aims, for one thing, to provoke us into developing new feminist STS practices for the natural sciences. As a feminist scientist, I would like to pause on the idea of cloning in order to create new modes of kinship and zones of proximity between feminist politics and molecular biology. Using the qualities of kinship and hylozoism, I want to think differently about the role and contributions of molecular actants and thereby "reconfigure" some current connections between feminist theory and biology. By suggesting that we use cloning as a feminist practice in the natural sciences, I aim to describe some of the complexity that surrounds a feminist when they find themselves as a knower in the biological sciences. This practice may allow the feminist scientist to address the dilemmas they face in the lab and turn these dilemmas into micropolitical actions.

Second, by developing subcloning into a feminist practice, my intention is to create movement through a form of strategic mimesis. Referring to Massumi's work and the politics of identity, Braidotti states that "strategic mimesis [is] a positive simulation that does not essentialize an original. The point is to aim at the transformative impact of one's political processes."[22] Following Stengers, I also believe that the trope of cloning "belongs to the present as a vector of becoming or an experience of thought—that is as a tool of diagnosis, creation and resistance."[23] The reason for posing this risky question, then, is to situate myself as a feminist scientist and to "turn" our attention to new lines of flight that can come with cloning practices.

Cloning Subcloning

Subcloning is an integral part of molecular biology research. In this technology the scientist begins with a gene or gene fragment of interest that is most likely obtained by amplifying DNA through PCR. The scientist then introduces or "clones" the gene of interest into a bacterial genome, typically bacteriophage lambda, which is also referred to as a plasmid or vector. Genetically engineered linear DNA is ligated into a plasmid and then placed back into a bacterial or yeast cell. As explained by New England Biolabs, a leading supplier of recombinant and enzyme reagents for life sciences research:

> Plasmid vectors allow the DNA of interest to be copied in large amounts and, often, provide the necessary control elements to be used to direct transcription and translation of the cloned DNA. As such, they have become the workhorse for many molecular methods, such as protein expression, gene expression studies, and functional analysis of biomolecules. During the cloning process, the ends of the DNA of interest and the vector have to be modified to make them compatible for joining through the action of a DNA ligase, recombinase, or in vivo DNA repair mechanism.[24]

Once the gene of interest has been isolated and then ligated (joined) to a vector, the next step in the molecular biology technique involves inserting the hybrid plasmid into "competent" *Escherichia coli* (*E. coli*) cells. Bacterial *E. coli* cells that are able to incorporate hybrid plasmids and successfully proliferate are referred to as being competent cells. Learning more about these competent *E. coli* bacterial cells is crucial to this chapter and to the process of becoming molecular.

To return to the subcloning procedure, the cloned vector is then inserted into the bacterial cells by heat shock and in this way *transforms* *E. coli* cells. Heat shocked and transformed *E. coli* cells are spread onto nutrient-rich LB plates and incubated overnight at 37°C to induce the growth of bacterial colonies. Of these colonies, about ten of possibly hundreds are selected for culture—a process whereby the plasmid DNA inserted within the transformed *E. coli* cells is amplified. As a colony of competent bacterial cells begins to replicate, it also replicates the genomic material of the foreign gene as if it were its own. The amplified plasmid DNA is then analyzed by restriction enzyme analysis and gel electrophoresis. In the final step of the molecular biology subcloning experiment, the amplified and cloned PCR product is analyzed by DNA sequencing. This is done to verify that the gene fragment of interest was obtained from the appropriate gene. Several biotechnology companies have optimized the technology of subcloning. To simplify the experimental process, these companies sell "cloning kits" to scientists. These standardized kits include inorganic reagents such as salt solutions and buffers as well as organic materials such as nucleotides, segments of DNA known as primers, a heat stable DNA polymerase enzyme isolated from the bacterium *Thermus acquaticus* called Taq polymerase, and *E. coli* bacterial cells—all required for subcloning.

While conducting my own research in molecular biology, I always used the Invitrogen® pCR™8/GW/TOPO® TA Cloning Kit.[25] The glossy and user-friendly manual provided with the Invitrogen® kits made the experimental procedures easier. The "Never Clone Alone" slogan posted in our lab (a sense of humor that can only be cultivated by scientists) was intended to put me at ease, constantly reminding me that I should not be alone (even though much of the time I was alone, conducting experiments late into the night). Indeed, I did not feel alone, for I had developed a close and personal relationship with these Invitrogen® kits, carrying the pocket-size manual close to my heart, in the breast pocket of my lab coat. The Invitrogen® cloning kit manual left a deep impression on my lab coat pocket and on me, making me yearn for a cloning technique of a different kind. The manual outlined the subcloning technique in these easy steps: (1) produce your PCR product; (2) perform the TOPO® Cloning Reaction; (3) transform into One Shot® Chemically Competent *E. coli*; (4) select and analyze colonies; and (5) choose a positive transformant and isolate plasmid DNA.[26]

As I subcloned, I took to heart Haraway's project of queering or mutating the modest witness and knew that I needed to subclone my way to something or somewhere completely different.[27] Describing her dream, Haraway wrote: "My modest witness cannot ever be simply oppositional. Rather, s/he is suspicious, implicated, knowing, ignorant, worried, and hopeful. Inside the net of stories, agencies, and instruments that constitute technoscience, s/he is committed to learning how to avoid both the narratives and the realities of the Net that threaten her world at the end of the Second Christian Millennium. S/he is seeking to learn and practice the mixed literacies and differential consciousness that are more faithful to the way the world, including the world of technoscience, actually works."[28]

While I subcloned, I always felt an urge to reproduce (perhaps by parthenogenesis) a clone of subcloning.[29] What follows is my attempt to clone subcloning into a practice that the feminist scientist can use to face dilemmas in their research and transform these dilemmas into desires for new scientific knowledges. The feminist practice that I yearn to produce resembles the five easy steps to molecular subcloning outlined in the manual provided with the Invitrogen® pCR™8/GW/TOPO® TA Cloning Kit. This feminist STS practice, which I refer to as Sub/FEM/cloning, consists of five steps: (1) isolate your dilemma; (2) ligate the dilemma to vectors of figuration; (3) transform the dilemma; (4) select and analyze new connections; and (5) collect your reconfigured dilemma. By pausing on each mode, the feminist scientist can move more freely through complex relations and enter into deeper zones of proximity between biology and feminist politics. Sub/FEM/cloning draws from Haraway's theory of situated knowledges, is intimately connected to Chela Sandoval's methodology of the oppressed, and is inspired by Barbara McClintock's work on transpositions and the cytogenetics of corn.[30]

Step 1: Isolate Your Dilemma

In the first step of Sub/FEM/cloning, the feminist must articulate that question or issue which is at the basis of their dilemma in the lab. This dilemma may stem from tensions based on the epistemologies, paradigms, language, methods, or tools that a feminist scientist may use in their practice of science. Underneath it all, the dilemma may be borne out of stolonic desires to create connections between disparate groups of knowers. This dilemma is at first destabilizing and disorienting, but in this act of articulation the feminist scientist isolates the stabilizing element of what Sandoval refers to in her methodology of the oppressed as *differential movement*. As an insider/outsider, marginalized-knower, hyphenated-cyborg creature, Sandoval argues that a "split-consciousness" allows one to "see from the dominant viewpoint as well as her own."[31] Differential movement can move the feminist scientist from the confines of her lab bench toward more complex micropolitical positions.

In my case, by articulating the question "Should feminists clone?" I isolated a dilemma that stabilized me in my differential movement as a feminist scientist working in a reproductive neuroendocrinology lab. During my PhD research I was interested in examining the regulation of hypothalamic neurons by gonadal steroids such as estrogen. I was also interested in examining the regulation of these neurons by the pineal hormone melatonin. I was most interested in searching for possible feedback control mechanisms in the hypothalamic-pituitary-gonadal (HPG) axis. In her book *The Woman in the Body*, Emily Martin provided ample evidence to suggest that in reproductive biology, the HPG axis had been depicted as a hierarchy, with the hypothalamus acting as the control center.[32]

Indeed, this understanding of the HPG axis has had a great impact on how scientists and doctors "manage" women's bodies, particularly in the treatment of menstruation and menopause. To counter this paradigm of a hierarchy and suggest the possibility of an alternative mechanism by which the HPG axis may function, such as feedback regulation, it first had to be established that estrogen and androgen receptors are expressed in specialized neurons of the hypothalamus. I was very excited to participate in one of my supervisor's research projects that investigated the possible neurological impacts of estrogen by examining its effects on an *in vitro* model of hypothalamic gonadotropin releasing hormone (GnRH) neurons. This research remains extremely relevant to women's reproductive health, particularly taking into consideration our lack of knowledge on the prolonged neurological effects of estrogen-based treatments such as contraceptives and hormone replacement therapies. In fact, this concern for women's reproductive health can easily be aligned with politics emerging from a molar position.

To do this research, however, I needed to look for the presence of an estrogen receptor gene and protein expression in these hypothalamic neurons—and to do this, I needed to use the molecular technology of subcloning. This was my dilemma. Even while I isolated my dilemma as a feminist scientist, the question of whether or not I should subclone, I also isolated total RNA from an *in vitro* cell line of GnRH neurons. I then synthesized first strand cDNA from total RNA using reverse transcriptase (RT) reactions. Using oligonucleotide primers (short sequences of DNA nucleotides) that were designed specifically for estrogen receptors and polymerase chain reactions (PCR), I amplified and obtained cDNA fragments of the estrogen receptor-alpha (ERα) and estrogen receptor-beta (ERβ) genes from these hypothalamic neurons.[33] The following steps describe the process I used to bring together my molar and molecular politics.

Step 2: Ligate the Dilemma to Vectors of Figuration

Similar to the bacteriophage plasmids used in subcloning, in the feminist practice of Sub/FEM/cloning, the dilemma must also be ligated to vectors, but in this case, to vectors of figuration.[34] The vectors of figuration produce a particular cartography or a map of relevant spaces, inhabited in my case by the material and discursive practices of cloning. As Rosalyn Diprose and Robyn Ferrell explain in their introduction to *Cartographies*:

> A map does not simply describe what is. A map does not only set up a grid which determines what can be found by selection or omission. Nor is it merely a series of lines inscribed on a previously blank surface. There is an alterity which provokes the desire to map, to contain and to represent. . . . The political reality of the changing map of the world, its allegiances, exclusions and oppression, is testament to cartography as a relevant metaphor. Mapping, as representation, is inextricably caught up in the material production of what it represents. In the metaphor of cartography, to draw a line is to produce a space, and the production of the space effects the line.[35]

As an insider-outsider and implicated knower, the alterity experienced by the feminist scientist is bound to manifest itself in the form of a dilemma brought on by inhabiting a space constructed by dominant and traditional scientific practices. It is true that, in order for feminism to change science, feminists need to "inhabit" the sciences. Yet how is this space to be made somewhat hospitable? To play on Wilson's use of the breach, I suggest that the feminist scientist must engage in the practice of "interior reconfiguration."[36] They must take hold of their dilemma and use it to produce an alternate cartography that better reflects their politics.

While describing the principle characteristics of a rhizome, Deleuze and Guattari comment on the relevance of maps, suggesting that "the map does not reproduce an unconscious closed upon itself; it constructs the unconscious. It fosters connections between fields. . . . It can be drawn on a wall, conceived as a work of art, constructed as a political action."[37] Cartography or mapmaking is hard work and constitutes the most time-intensive mode in the feminist practice of Sub/FEM/cloning. These acts of ligation can be thought of as being driven by what Sandoval has called *democratics*, the act of imagining social justice and positive change, and bringing into closer proximity particles that would otherwise wander without collision.[38] So the vectors of figuration that a feminist scientist encounters in order to create a new map are not simply figurative ways of thinking. This work is not destined to be muddled by relativism or idiosyncratic musings. Rather, these vectors of figuration provide a material mapping, allowing one to think with specificity.

For instance, in her later writings, it is through her practice of figuration that Haraway expanded on her model of situated knowledges. Using the idea of *stem cells* and *sticky threads*, Haraway states: "Objects like the fetus, chip/computer, gene, race, ecosystem, brain, database, and bomb are stem cells of the technoscientific body. Each of these curious objects is a recent construct or material-semiotic 'object of knowledge,' forged by heterogeneous practices in the furnaces of technoscience. . . . [O]ut of each of these nodes or stem cells, sticky threads lead to every nook and cranny of the world. Which threads to follow is an analytical, imaginative, physical, and political choice."[39] The marginalized-knower-feminist-scientist, also known as a cyborg in some circles, must recognize the value of their own specific analytical, imaginative, physical, and political choices in why and how they conduct their science. According to Haraway, the stem cells and sticky threads are the embodied consciousness of any given situated knowledge.

Embodied consciousness? Cyborgs? Allow me to flesh out this mess of stem cells, sticky threads, and ideas. As a six-year-old, I recall an incident that might very well have been my very first lived experience as a cyborg, or at least was a formative moment for my own personal "cyborg politics." Describing this incident may help to explain my difficult but intimate relationship with science and technology. It was Toronto in the late 1970s and I was walking home from school sometime in the early part of spring when the sidewalks were clear, but there were still patches of snow here or there. I was almost home when an older student started throwing snowballs at me from across the road. As he threw each ball of snow mixed with rocks, he yelled the racial slur "Paki!"[40] At the time, as a six-year-old, I did not know what "Paki" meant and I did not know why he decided to throw snowballs at me. All I knew was that I was going to be able to run away very quickly from this bully. For what I knew, and he did not, was that I was *bionic*. Sure enough, having on my Bionic Woman running shoes helped, and so I was able to motor my way home.

Visions of biotechnological futures saved me that day. Very early on in my makings as a feminist scientist-cum-cyborg, I realized that I was going to face difficulties in being both "bionic" and "brown"—two of the many stem cells, or material-semiotic "objects of knowledge" that have since come to form my conception of the technoscientific body. Which figurations a feminist scientist brings into closer proximity depends on the stem cells and sticky threads that have come together to bring forward their dilemma. So being "bionic" and "brown" can be thought of as two stem cells that I acknowledge as playing a part in constructing my reality. As Haraway explains it, each stem cell is comprised of a "knot of knowledge-making practices" formed by such sticky threads as "industry and commerce, popular culture, social struggles, psychoanalytic formations, bodily histories," and more.[41]

In my scientific work, after the estrogen receptor gene fragments had been isolated from the cDNA obtained from *in vitro* GnRH hypothalamic neurons, these PCR products were electrophoresed in an agarose gel, stained with a dye called ethidium bromide and visualized under UV light.[42] The DNA fragments were then isolated and ligated to the pCR2.1-TOPO cloning vector (circular bacterial plasmid DNA) provided by the biotechnology company Invitrogen®.[43] While I was conducting this most problematic recombination reaction, in my own Sub/FEM/cloning experiment I was forced to ligate my dilemma to vectors of figuration. Through this process I became aware that my dilemma must be combined and connected to broader contexts. The vectors of figuration that I encountered while trying to make some meaning out of my dilemma of subcloning were obtained from my own analytical, imaginative, political, and physical senses of being bionic and being brown at the same time. There are a number of possible figurations to explore, but I will restrict the discussion of my particular dilemma in the context of three figurations: (1) Shulamith Firestone, (2) Rajasthani prints, and (3) Superman.

Step 3: Transform the Dilemma

In the Sub/FEM/cloning transformation process, the dilemma that has been ligated to vectors of figuration must be used to create a transformation. In this mode of a Sub/FEM/cloning experiment, the feminist scientist must take their isolated and ligated dilemma and move toward new scientific knowledges. Our hope lies with becoming like the competent *E. coli* bacteria, by gathering our strengths and harnessing our abilities to transform. This transformation in the feminist scientist reconfigures their outlook on the science that they practice. This new position can allow the feminist scientist to address their original dilemma and step in a direction that creates movement.

In her account of the life and work of Barbara McClintock, Evelyn Fox Keller brought to our attention the extraordinary story of a scientist who worked on the genetics of corn. By paying such close attention to the details of McClintock's life and her approach to scientific research, Keller revealed much more than this, however. Throughout her life McClintock had many scientific accomplishments, one of which was the discovery of transpositions—the movement of genetic elements spontaneously from one site to another. As described in a previous chapter, McClintock's work on transpositions came out of her scientific approach of developing a "feeling for the organism." McClintock describes how, while analyzing the chromosomes of maize through the eye of the microscope, she would travel deep into the cell and find herself in and among the chromosomes, almost becoming imperceptible. Keller describes this level of association with an organism as a kind of "intimate knowledge."[44] This expression of proximity most vividly exemplifies a microphysiology of desire.

As McClintock observed the phenomenon of transpositions, she came to realize that transpositions must happen, that organisms are not stable, and that DNA and chromosomal structures get mixed up. "In McClintock's system," Keller writes, "the controlling elements did not correspond to stable loci on the chromosome—they moved. In fact, this capacity to change position, transposition as she called it, was itself a property that could be controlled by regulator, or activator genes . . . no one was ready to believe that, under certain circumstances, the normal DNA of a cell could rearrange itself."[45] Referring to McClintock's work in relation to DNA and the function of transpositions, Keller continues: "Perhaps the future will show that its internal complexity is such as to enable it not only to program the life cycle of the organism, with fidelity to past and future generations, but also to reprogram itself when exposed to sufficient environmental stress—thereby effecting a kind of 'learning' from the organism's experience."[46]

Through this transformation step in the feminist practice of Sub/FEM/cloning, the feminist scientist can attempt a transposition or "reprogramming" of their feminist politics. In the context of my dilemma, for instance, I was able to acknowledge that as feminists, we have been exposed to many new stresses in the past twenty to thirty years. We have been exposed to new biotechnologies, reproductive technologies, and molecular technologies, and these experiences have changed the way we live. The figurations that I describe below are meant only to serve as examples and are articulated through my own experiences of alterity. I share these thoughts with the hope of providing some snapshots of "the political reality of the changing map of the world, its allegiances, exclusions and oppression."[47]

While working in a reproductive biology lab, I knew that strange things were happening constantly around me, but I was determined to go deeper into the science that I practiced until I was able to find the molecular practices that would show me how to transform my dilemma. I studied in more detail what was known about the HPG axis and spent a great deal of time working closely with an *in vitro* cell line of hypothalamic neurons. It was here that I found a new space emerging for my feminist politics. I confess that I isolated fragments of estrogen receptor genes from hypothalamic neurons. I moved these DNA gene fragments, not spontaneously but rather through ligation techniques, into bacterial plasmids. I transformed competent *E. coli* cells with these plasmids and used their cell machinery in order to amplify estrogen receptor DNA. Each transformation reaction required several agar plates, a vial of *E. coli* cells, and a specialized media containing tryptone, yeast extract, glucose, and several salts.[48] The ligated vectors were added to the *E. coli* cells, placed on ice, heat shocked, and transferred onto ice again. The heat-shock process allowed the ligated vectors to enter through the membranes of the competent *E. coli* cells. A small amount of the transformed bacteria was spread over selective plates of agar and incubated overnight at 37°C. Just as the bacterial cells needed to incubate to transform, as a feminist scientist I also had to allow vectors of figuration to incubate for a while to form a rich array of connections.

Step 4: Select and Analyze New Connections

An efficient cloning reaction will produce nearly one thousand colonies of bacteria per transformation reaction. The ligation of the DNA fragment causes an interruption in a particular gene of the cloning vector. After a night of incubation, the agar plate containing the transformed *E. coli* cells will appear to have white colonies as well as blue colonies. Each white colony represents a single *E. coli* cell that was successfully transformed by a ligated vector and was able to multiply. If no ligation takes place and a particular marker gene on the plasmid is kept intact, this gene codes for a protein that can be made to react with another substance, turning the bacterial colony blue. Out of the nearly one thousand colonies that can grow on an agar plate, the scientist usually chooses a few of the white colonies or "positive clones" to examine. The colony of transformed bacterial cells is selected and amplified further in a broth of nutrient media. In my own experiments I often selected up to ten colonies for analysis. Using a sterile wire tip, I scooped up cells from the white colonies and added them to test tubes that were full of nutrient media. The transformed *E. coli* cells were left in a 37°C shaker for several hours. This constant movement and the connections formed by the collision of cells allowed the transformed *E. coli* cells to multiply even further. Having already ligated my dilemma to vectors of figuration, it is now time to select and tease out some of the connections created by my own Sub/FEM/cloning reaction.

Figuration 1: Shulamith Firestone

After the birth of my second child, I found myself in a curious predicament. I had to return to my teaching responsibilities after just six weeks of maternity leave (which was actually categorized by the institution as a disability leave). It was necessary for me to pump breast milk before and after teaching my three-hour upper-division women's studies course, "Science and Technology in Women's Lives." While I pumped in my office, the pile of books on my desk that I had been meaning to read stared at me, so I decided to multitask and catch up on some reading. At the top of the pile was the newly released 2003 edition of Shulamith Firestone's *The Dialectic of Sex*. First published in 1970 and considered an essential text of second-wave feminism, in this text Firestone put forward her feminist theory of politics.

"In the historical interpretation we have espoused," she wrote, "feminism is the inevitable female response to the development of a technology capable of freeing women from the tyranny of their sexual-reproductive roles—both the fundamental biological condition itself, and the sexual class system built upon, and reinforcing, this biological condition."[49]

Firestone was referring here to the advent of hormonal contraceptives and their potential role in the liberation of women. She was very clear in stating what she saw as the single most influential reason for women's oppression: their capacity for reproduction. She went as far as saying that "pregnancy is barbaric" and was in favor of using the power of technology to rid women of the burden of reproduction.[50] Sitting in my office, I used the technology of a breast pump, pumping with one hand and holding a book in the other. It seemed to me that it was not pregnancy that was barbaric, but rather the circumstances under which I was expected to function. In the state university system in which I was an employee at the time, my tenure clock was not stopped during my brief maternity leave; furthermore, I was expected to be able to lactate and teach only six weeks after giving birth. There was quite obviously an assumption made on Firestone's part, and echoed by other feminists at the time, that *all* women find motherhood to be a burden and experience oppression in the same way because of their capacity to reproduce. This assumption ignored the possibility that some women are not in a position to be able to afford to reproduce or are in fact forced into not reproducing. The erasure of issues pertaining to the intersections of race, class, disability, and sexuality and the bodily histories of marginalized others within this era of feminism became quite apparent to me while reading Firestone's work.

Despite this, most fascinating was Firestone's accuracy in predicting the direction that artificial reproduction would take from the period in which she was writing in the late 1960s. She even predicted the technology of cloning and was excited at its potential to liberate women. "[As] recently as five years ago," she wrote, "Professor F. C. Steward of Cornell discovered a process called 'cloning': by placing a single carrot cell in a rotating nutrient he was able to grow a whole sheet of identical carrot cells, from which he eventually recreated [*sic*] the same carrot. The understanding of a similar process for more developed animal cells, were it to slip out—as did experiments with 'mind-expanding' drugs—could have some awesome implications. Or, again, imagine parthenogenesis, virgin birth, as practiced by the greenfly, actually applied to human fertility."[51] Firestone was not suggesting the use of a technology such as cloning without further examination. She was fully aware that in the wrong hands artificial reproduction would be dangerous. At the same time, she felt that artificial reproduction was inevitable. She believed that in order to deal with the inevitable, we would have to create a new culture based on a "radical redefinition of human relationships."[52] This radical redefinition, she believed, would force societies to destroy current class systems and ideas of family. Firestone did not think that artificial reproduction was inherently dehumanizing, and she believed in the potential of this technology in freeing women from their biology. At the end of *The Dialectic of Sex*, she lays down a "list of demands" for a feminist revolution. "The freeing of women from the tyranny of reproduction by *every means possible*," the first demand insisted, "and the diffusion of the child-rearing role to the society as a whole, men as well as women."[53]

Beginning with the ligation of my dilemma to the figuration of Shulamith Firestone is a difficult maneuver. It reveals several hidden patterns of molar politics that are present within a great deal of contemporary feminist discourse. For example, Firestone pointed out that there is a series of feminist ethical positions that fit into one another beginning with the belief that technology is inherently evil, followed by the belief that technology is bad for women, and lastly, that artificial reproduction is dehumanizing. She argued that artificial reproduction is not inherently dehumanizing by forcing us to reconsider what we know and believe to be "natural" and question our ties to a certain mode of reproduction. Her work challenges us to reconsider our relationship with, or in more accurate terms, our distrust of technology. Extending Firestone's assertion, many lesbian and feminist science fiction authors actually address the oppression of women in our societies by creating worlds where women and/or female bodies are no longer solely responsible for pregnancy, childbirth, breastfeeding, and childrearing. A common theme running throughout these science fiction and utopian novels has been that of supporting alternate models of reproduction, and these models always involve some form of genetic manipulation. Parthenogenesis and cloning are very popular in these works.[54] For many feminists the belief is that only men would want to design and control technologies related to reproduction. The figuration of Firestone exposes the idea that some women may also support technologies such as cloning. Similarly, while commenting on some of the popular biological techniques used for reproduction in science fiction novels, feminist science fiction writer Pamela Sargent stated, "What is the extent of possible biological change? It can involve new ways of reproducing ourselves, a use of techniques such as cloning, ectogenesis (the use of an artificial womb), in vitro or 'test-tube' fertilization, hybridization of animal species and humans, and others. . . . Biological change could in time affect our notions of what a human being *is*."[55]

In the context of my dilemma of subcloning, the figuration of Firestone exposes several feminist projects such as those based on social struggles against traditional family structures. In order to gain entrance into a market economy, women have been forced into living as commodities by way of their reproductive potential. The most challenging connection that emerges from the ligation of my dilemma with this figuration, however, concerns the question of what it means to be "human." An analysis of Firestone's work and the lesbian and feminist science fiction inspired by her work also forces us to ask, What is natural? By problematizing the validity of a "natural" mode of reproduction and imagining biological change to the extent that we can no longer easily define what a human *is* (something that has already occurred through the use of transgenic technologies and bionic woman in my case), this figuration forces us to reconsider what it means to be human. A revised notion of "the human," or a posthumanist understanding of what it is to be human, may permit us to imagine the answer to this question to be a flexible amalgam of altered bodies, senses, and subjectivities.

Figuration 2: Rajasthani Prints

I include this figuration as part of my own analytical and imaginative mapping of a social reality because of its relevance to reproductive technologies and its deep impact on me as a brown child and later as a young brown adult visiting Rajasthan, India. Growing up in a home with parents who had immigrated to Canada from India, I was surrounded by explosions of color and texture from various Indian artworks and sculptures that decorated our home. One such work was a Rajasthani print displayed as a central piece in our living room, above the sofa. As a child, I would climb onto the sofa and carefully examine each minute detail of this print. What fascinated me the most was the intricate design that served as a border. The hand-drawn border was comprised of a repetitive series of women dressed in identical clothing, with identical expressions on each face. As a child, I tried to find a woman drawn along that border that did not match, but I never could. The women depicted in the print were identical; they were clones.

Years later, as a once-removed NRI (nonresident Indian) on a trip to India, I had the opportunity to visit Rajasthan. It was 1994—the same year as the International Conference on Population and Development (ICPD) of the United Nations Population Fund in Cairo. In one village I saw an "epidemic" of small, square patches pasted onto the arms of poorer women walking in the streets. These real women in Rajasthan were not clones, but each bore the mark of what I think was a clinical trial for a new contraceptive on the upper outer regions of their left arms.[56] On this trip I was hosted by a generous woman and her daughter, who was fiercely independent. When I asked about the women in the streets whom I had seen participating in what I thought might be a contraceptive trial, I was made acutely aware of my position as a "Western" feminist in this context. As a diasporic Indian, I recalled the incident that I had experienced as a six-year-old—having been called a "Paki"—and realized that this was not the first time I was made to feel as though I did not belong to the space I inhabited. In any case, despite the anti-Malthusian arguments put forward by many local and national Indian feminist organizations, I was informed by my younger host that state-sanctioned family-planning incentive programs were beneficial to the poor and scheduled castes, and that the regulation and policing of reproductive bodies was necessary to better serve national economic interests.[57] Incidentally, this was also the year that Rajasthan began to implement a "two-child norm" policy for government employees; to date, this policy is still in effect.[58]

The women who were participating in what might have been a contraceptive trial in Rajasthan at the time are like many other women around the world, and not just those in so-called developing countries. All of these organic bodies function as commodities in the institutions of science, medicine, and the state. In many ways, women already exist as clones as their reproductive body parts are disassembled, traded, and reassembled in the technological production lines of scientific and pharmaceutical research. As a result of our expendability, women's whole bodies as well as individual reproductive body parts have been used as test subjects in the name of scientific progress. When women are the subjects of scientific study, they generally come to exist through a process of standardization. The notion that "women" can exist as a single category of organic beings who contain reproductive body parts that simultaneously have specific yet transferable technical capabilities, is in itself a molar position that forces individual identities into a cloned existence. Much like the repetitive pattern of side-glancing women who served as a border in the Rajasthani print of my childhood memories, many women around the world already exist as clones.

Postcolonial and decolonial perspectives can help us to see that recent formulations of biocapital have created new forms of kinship between women, animals, and plants. They all share the experiences of forced modification and commodification. In India, where clinical trials for new drugs and contraceptives have been conducted, the Green Revolution also occurred, whereby several multinational corporations with the intent of colonization—albeit of plants and not of humans—entered into the country. Around the world, parallel patterns can be drawn between the production of reproductively modified women and the production of genetically modified plants and animals. Both are organic material necessary for progress in a culture of cloning. The ligation of my dilemma to the figuration of Rajasthani prints helps to make apparent the connections between women, reproductive biology research, pharmaceuticals, plants, genetic engineering, and multinational corporations. These newly formed sticky threads force me to trouble my practice of subcloning and consider more closely the impact that cloning technologies have had on the bodies of women of color, and how often they have served as material test subjects.

Figuration 3: Superman

This superhero first appeared in a comic strip in 1933, but since then, an endless number of incarnations have appeared on television, film, and even radio. The animated action hero easily materialized into a living and breathing character, first as George Reeves in the 1950s television series and later as Christopher Reeve, who took on the iconic role of Superman in the 1978 movie (just around the same time that I would have been at the peak of my brown girl bionic fierceness). This latter materialization was quite convincing, to the extent that even today the identity of Superman is synonymous with the late actor. In the mid-1990s, when Christopher Reeve fell off a horse and severely injured his spine, part of me believed he would be the first human to walk again after a serious spinal cord trauma. After all, he was Super(hu)man. However, Reeve did not walk again. Following his accident, he became an activist and advocate for medical research to help people living with paralysis. Just days before his death in October 2004, one month before the US presidential elections, Reeve made his last public appeal on television for the support of stem cell research. From the likes of the actor Michael J. Fox, to the wife and son of the late Ronald Regan, Superman and his (super)friends from Hollywood, California, have taken their plight from the Hall of Justice to another legislative assembly, the US Congress. In as heroic a gesture as battling the Legion of Doom, superhero celebrity figures in the United States have made it their mission to garner support for human stem cell research, also referred to as human therapeutic cloning.

Unlike reproductive cloning, human therapeutic cloning does not attempt to reproduce an identical human. Rather, the "purpose of therapeutic cloning is to generate and direct the differentiation of patient-specific cell lines" that can be used for personalized medicine and involves the "transfer of nuclear material isolated from a somatic cell into an enucleated oocyte in the goal of deriving embryonic cell lines with the same genome as the nuclear donor."[59] Not surprisingly, a great deal of controversy has accompanied this biotechnology. Although in theory therapeutic cloning research can be conducted using either adult or embryonic stem cells, much of the controversy has been over the creation and use of embryonic stem cell lines. In a script that almost plays out like a superhero versus villain drama, the controversy continues to rage, with emotions flaring high for both those opposed to the technology on the basis that it is immoral and unethical, and for those who support the technology for its potential to cure diseases, as scientists have promised.

To summarize some key scenes from this stem cell drama as it has played out in the United States, we can look at the battle between the federal government and the state of California.[60] This story begins on August 9, 2001, when President George W. Bush announced that no further federal funds would be used to support human embryonic stem cell research in the United States.[61] He made this decision on the grounds that stem cells obtained from human embryos at the early stages of development constituted the unethical treatment of human beings. Scientists argued that in order to proceed with research in a meaningful way, they needed to create new human embryonic stem cell lines and receive support, financial and otherwise, from the federal government.[62] No doubt with a keen eye to the lucrative potentials of this biotechnology, the state of California (a vortex in the universe where vectors of figuration perpetually hyperimplode) responded by passing a state proposition in 2004 supporting the issue of a $3 billion bond to fund human embryonic stem cell research for ten years at a staggering amount of $300 million a year.[63]

Enter stage left another super(hu)man: Arnold Schwarzenegger, then Republican governor of California but also once a cyborg action hero. If nothing else, this appearance makes clear that Hollywood superheroes support human therapeutic cloning research. Putting himself at odds with the Bush administration, Schwarzenegger endorsed the $3 billion bond measure in part to boost California's biotechnology industry.[64] In response, Bush exercised his first veto as president in 2006, rejecting legislation that would have increased the annual $25 million of federal funding for embryonic stem cell research. In 2007, California's $3 billion state bond program (referred to as the California Institute for Regenerative Medicine) approved the distribution of $45 million for embryonic stem cell research, promising to fund an additional $80 million to established stem cell researchers.[65] Two years later, in March 2009, President Barack Obama issued an executive order "removing barriers to responsible scientific research involving human stem cells," effectively lifting the ban on human embryonic stem cell research that the Bush administration had put in place.[66]

The back and forth of this script is almost comedic, yet it should come as no surprise that in 2017 conservative representatives once again urged for stronger restrictions on human embryonic research and are calling upon the current US president to fire Dr. Francis Collins, director of the National Institutes of Health, for his role in moving this research forward.[67] At the same time, ten years after starting up its initial stem cell program, California is ramping up yet again for an aggressive campaign on an upcoming state ballot (at the time of writing) that would put a funding measure in place to continue stem cell research. Although cures for Alzheimer's, Parkinson's, spinal cord injuries, and other diseases that the program had envisioned have not yet been delivered, for many the recent success of treating a young girl from Corona, California, who suffers from Severe Combined Immunodeficiency or "bubble baby disease" is sufficient proof that the research is highly beneficial and desirable.[68]

Most governments around the world have placed a ban on human reproductive cloning, but their position on human therapeutic cloning varies. Those opposed to human therapeutic cloning research typically argue from the moral position that stem cell research violates the dignity of human beings. They are concerned with the rights of the unborn child, much like the line held by pro-life advocates in the abortion debate and argued in some cases with a similar evangelical fervor. Also of interest are arguments put forward by supporters of stem cell research, which includes several Hollywood stars. In the name of supporting our inevitable biotechnological destiny, stem cell research enthusiasts argue that this scientific research must be allowed to take place. In a typical humanistic vein, the argument put forth is that by denying stem cell research, not only are we depriving the quality of life for those humans who are currently suffering from diseases, we are also denying the "natural" process of human discovery, thereby denying human progress, and ultimately human life.

Those humans left off both the moralistic and humanistic radars of those engaged in the stem cell debate are the women whose bodies upon which this technology is to be developed. The discourses produced by both sides of the stem cell research debate fail to address any concern for the women whose reproductive parts—from ova to umbilical cords—are necessary for the scientific research. The attitude on both sides of the debate is based on a popular consciousness that allows us to believe that women are merely resources for biological material and that this technology can and will be developed on women's bodies. For instance, in light of legislation that would allow Australian scientists to move ahead with therapeutic cloning, Catherine Waldby pointed out concerns regarding the trafficking of ova and the exploitation of poor women around the world to support the rapidly developing market for human eggs.[69] Waldby's comments were dismissed by an Australian stem cell researcher who, failing to see her point, stated: "Waldby's concerns are of no direct relevance to Australia."[70]

This figuration presents yet another tension. In a most uncomfortable move, many feminists in the United States have found themselves in a strange alliance with the religious right and with social conservatives by voicing their opposition to therapeutic and reproductive cloning research. A molar politics based on the assertion that human life is more important than animal life acts as a corollary to the established hierarchical frameworks in which our society operates. The political and religious right in the United States support patriarchy within this hierarchical structure, granting men a certain status and placing women below men. Following this logic, nonhuman animals fall below women, although perhaps not far behind. Many feminists, who are opposed to the patriarchal elements within this hierarchy, end up supporting this structure by also placing nonhuman animals below themselves, not recognizing the importance of kinship and hylozism. This allows one to oppose human stem cell research while ignoring or even supporting the cloning and/or genetic manipulation of animals for medical research, such as in breast cancer research. Feminists as well as members of the political and religious right, who are opposed to reproductive technologies, both conveniently ignore or fail to realize the decreased value they have placed on lives that are not human.

In March 2005 the United Nations General Assembly voted in favor of a nonbinding ban on all human cloning, which in less clear language also includes therapeutic human cloning.[71] In this UN ban on reproductive cloning, however, only human reproductive cloning is specified, therefore permitting reproductive and therapeutic cloning research on such animals as mice, cats, sheep, horses, monkeys, and cows. The United Kingdom and South Korean governments, for instance, have banned human reproductive cloning but do support animal reproductive cloning. As such, these governments have provided federal funds to scientists who have produced significant developments in animal cloning, the most famous of which includes Dolly the sheep.

Although much is known about the science behind the birth and death of Dolly the sheep, what is perhaps less well known is the namesake behind this first mammal that was produced through reproductive cloning. Dolly was named after the American country singer Dolly Parton because of a strange kinship relation based on mammary glands. The association between the two may not be immediately clear to many of us but is apparently based on the fact that the cloned sheep was derived from the nuclear transfer of an adult somatic mammary gland cell, and that the superstar Dolly Parton has large breasts. In another strange kinship alliance made possible through reproductive cloning, grieving pet owners can now replace a lost pet by purchasing a reproductively cloned animal. Once again, from the vortex of stem cells and sticky threads that is California, the company Genetic Savings and Clone Inc., which was located in Sausalito, California, was the first to offer such services. The company closed its doors at the end of 2006, claiming that there was not enough demand for their product but directed anyone who was still interested in freezing their pet's DNA for future cloning possibilities to the Texas-based company ViaGen.[72] As of 2017, ViaGen offered several services, including genetic preservation, reproductive cloning, and express tissue banking. According to their website, the "total cost of dog cloning is $50,000 [USD]" and the "total cost of cat cloning is $25,000 [USD]."[73]

The figurations of Shulamith Firestone, Rajasthani prints, and Superman create cartographic connections between a lab technique and new micropolitical positions. After placing a spotlight on the entanglements between biotechnology, popular culture, art, women, and animals, I conclude with the last mode of the Sub/FEM/cloning experiment.

Step 5: Collect Your Reconfigured Dilemma

The reason that Sub/FEM/cloning may actually work is that the feminist scientist can use this reflexive feminist practice for the natural sciences to arrive at Sandoval's *differential consciousness* without ever having to leave their lab bench.[74] As a hyphenated creature, living on the margins, the feminist scientist will have already created a space for their survival and feminist imaginings in the more confined spaces of the scientific institution in which they operate. This space, inhabited by the marginalized and existing within a space that is marginalizing, is what some might refer to as cyberspace.[75] A *differential consciousness* allows the feminist scientist to enter into this cyberspace, where they can conduct their experiments and create new scientific knowledge.

During my own escapes to the space I occupied while a graduate student working in a lab, I was first confronted with the monstrous question, Should feminists clone? It has since occurred to me that as a feminist-scientist type of hyphenated creature, the questions that I pose might not be monstrous at all but that I may in fact be the monster, and thus my attraction to these types of questions.[76] Drawing from Haraway, Sandoval describes a monster as "a creature who lives in both 'social reality' and 'fiction' and who performs and speaks in a 'middle voice' that is forged in the amalgam of technology and biology—a cyborg poet."[77] If I am a monster, or aspiring to become one, speaking in a middle voice places me in a favorable position to meta-ideologize.[78]

In an attempt to meta-ideologize, it becomes sensible and almost necessary to suggest that subcloning should become a feminist practice in the sciences. Describing *differential consciousness*, Sandoval explains that the "manipulation of ideology" is a necessary skill for the survival of the marginalized.[79] Which ideologies are manipulated, and in which direction to proceed, depends on one's context. "Such a differential force, when understood as a technical, political, aesthetic, and ethical practice," Sandoval states, "allows one to chart out the positions available and the directions to move in a larger social totality. The effectivity of this cultural mapping depends on its practitioner's continuing and transformative relationship to the social totality. Readings of this shifting totality will determine the interventions—the tactics, ideologies, and discourses that the practitioner chooses in order to pursue a greater good, beginning with the citizen-subject's own survival."[80]

The reason the feminist scientist is faced with a dilemma in the first place is in part due to their intimate relationship with science and technology. Is it possible for the feminist scientist to use their micropolitics to move toward a greater good and develop new knowledge without giving up their connection to science and technology? It is imperative that the feminist scientist continues to have this relationship, though strained, with the very science and technology that they wish to transform. My own intimacy with subcloning determined my intervention in the pursuit of creating new biological knowledges of the body. After ligating my question to vectors of figuration such as Shulamith Firestone, Rajasthani prints, and Superman, my dilemma was reconfigured. I became aware of recurring themes in the politics of cloning and this motivated a transformation in my research or a desire for movement in what already counted as established knowledge in the field. The molecular biology technique of subcloning allowed the isolation, the amplification, and finally the DNA sequencing of gene fragments that were of most interest to my research.

Subcloning allowed me to demonstrate the expression of estrogen receptor genes in hypothalamic neurons. The possible expression of estrogen receptors and the direct action of estrogen on hypothalamic neurons in this location of the brain had been dismissed prior to my research.[81] The significance of this finding, therefore, was that it contributed to new research on the HPG axis by providing evidence that the hypothalamus and gonads may interact through a series of feedback mechanisms rather than a hierarchical structure. Most important, I was able to bring molar and molecular politics to work together as my research contributions helped bring attention to the possible neurological effects of estrogen-based drugs, contraceptives, and hormone replacement therapies. I created a feminist account of the brain that was articulated through molecular biology.

As a feminist scientist, I have always been extremely appreciative of feminist critiques of science, but I have yearned to go beyond these critiques. During my PhD research, I wanted to engage with the biological sciences to produce a new feminist account of genes, hormones, receptors, and neurons. I had an opportunity to address my concerns for reproductive justice issues at a molecular level, which is why I subcloned. In her discussion of practices that can be used to approach and engage the sciences, Stengers reminds us that feminists may have to take the risk of "giving up the position of a judge."[82] To develop her ecology of practices, Stengers draws on Deleuze and his idea of "thinking par le milieu" and predicts some of the difficulties that may result from attempting to move from a majoritarian (molar) way of thinking to this minoritarian (molecular) thinking.[83] She suggests:

> I would thus claim that an important divergence between thinking in a major or in a minor key may well concern the relation between thinking and what we may call, in each case, ethics. The need and power to define a central stage is obviously determined by a political and also an ethical, project. . . . The problem, for me, is that such a characterization leads to identify the thinker's task as one of enlightenment, a critical and deconstructive enlightenment aiming to subvert the hegemonic languages and social structures, in order to free the constituent power which by right belongs to the multitude only. This is ethics in a major key since it implies and means to enact the great convergence between Truth and Freedom.[84]

Stengers explains in her ecology of practices that the difference between technology and the power of Truth is an ethical one, whereby technology is accompanied by a "sense of responsibility that Truth permits us to escape."[85] By engaging with the micropolitics of cloning, I have endeavored to work on the side of technology and not Truth.

Those who would answer my question regarding cloning with an immediate and resounding "no" may be doing so from a molar position that is bound to Truth. Molecular politics, however, can encourage us to engage with "a world that is technologically and globally mediated."[86] The purpose of having feminists enter into the sciences is not simply to keep the "women in science pipeline" piping. The goal instead is to create new biological knowledge that feminists desire. We want feminists to enter into the biological sciences. But once they are there, what should they do? Should they avoid the science and technologies that comprise the political economies of our time, or should they set up rebellion camps from within? Like it or not, encouraging feminists to enter into the biological sciences to produce new knowledges involves supporting them as they use the technologies that are crucial to their discipline.[87] As Braidotti warns: "What looks from one angle therefore as a potential threat of contamination of the minorities by the dominant norm or standard, from another appears instead as active resistance and innovation. This is not relativism, but the politics of location."[88]

I certainly would not have articulated my position this way while I was doing my PhD, but looking back now, I was driven by a sense that molecular politics were just as crucial as molar politics. My molecular position was absolutely necessary if I was to relate to the world around me as a feminist scientist in what I saw as being more productive ways. Subcloning became my transporting machine for spreading feminist contagion within the science that I practiced. What if the question "Should feminists clone?" were posed one last time? I hope I have made the case that if as feminists we are willing to get our hands dirty, and if we are prepared to extend kinship and hylozoic qualities to animals, plants, and even machines that have been created by the culture of cloning, it may be time to consider, even if just for a moment, that the answer to my monstrous question might be "yes."

5

In Vitro Incubations

> Humans are part of the world-body space in its dynamic structuration. Does this mean that humans have no responsibility for the outcomes of specific practices? . . . [D]oes that mean that human subjects are merely pawns in the game of life, victims of the same practices that produce the phenomena being investigated?
>
> —KAREN BARAD

> The heart of the project of philosophy as Deleuze conceives it . . . show[s] that going "beyond" the human condition does not entail leaving the "human" behind, but rather aims to broaden the horizon of its experience.
>
> —KEITH ANSELL PEARSON

Years ago, while preparing to split nearly confluent plates of immortalized mouse hypothalamic neurons, I had an experience in the lab that gave me pause. That pause has lingered with me ever since. It took place during a routine cell subculturing procedure, while I sat in front of a cold sterile fume hood with my latex-gloved hands placed on the door handle of a warm 37°C CO_2 incubator. Something happened in my otherwise unremarkable lab routine: I felt guilty for the shock in temperature I was about to inflict on unsuspecting neurons growing inside a warm bath of media and nutrients. I held off on opening the incubator door for a few labored seconds, then immediately considered the absurdity of the moment. I asked myself, What would these *in vitro* neurons think if they knew what I was about to do?

I had caught myself not questioning whether those neuronal cells growing *in vitro* were alive, or whether they could think, but rather, assuming that these neurons were living and already knew how to know, I was concerned with what their *response* would be in the moment that was about to unfold. I didn't quite see myself as a human pawn or a victim entangled in this phenomenon. Instead, I wondered about the lives of these neurons and the complex set of events that had come together at that moment to create an ontological crisis for me out of what was otherwise the mundane molecular biology practice of passaging cells. Only by spending time in a lab, and by using the practices of molecular biology, have I learned to generate such questions about nonhuman life and biology. This in turn has prompted me to consider an ethics of matter.

The *in vitro* lab protocol known as splitting, subculturing, or passaging allows the molecular biologist to keep cell lines alive for use in experimentation. However, unless one is willing and able to deal with an entire population of multiplying cells in culture, it also requires getting rid of, or killing, a vast number of these cells. In my case, once the cell line I was using was established, approximately 75–80 percent of the growing cells were discarded during any given cell passage procedure if they were not being used for experimentation; the remaining cells were placed in new plates with fresh media, fetal bovine serum, and other goodies to stimulate cell growth and division. Hannah Landecker has provided a rich analysis of how cells actually became technologies; she argues "that the history of cell cultivation is the history of an approach to living matter."[1] As part of her methodology of examining this history, Landecker places an emphasis on the importance of practice and the material basis of research, stating:

> Attention to the things people work with in experiments and to the ways they attempt to stabilize living objects such as the cell for scientific study has allowed historians and anthropologists to address the conditions under which scientific novelty is produced. Looking closely at the routine or infrastructural conditions that constantly allow the production of new things is a method for getting around having to explain all scientific developments as a "paradigm-ordered or theory-driven activity." In other words, the scientist does not have to think of it first, and

> act on the biological thing accordingly; change can arise from the objects and practices of experimentation themselves—how cells are kept, watched, represented, manipulated, and how they react and adapt to their technical milieu.[2]

Landecker's methodology deeply resonates with the emphasis I have placed on learning from experimental science and developing a practice-oriented feminist STS. However, Landecker also notes that it is by observing the everyday experimental activities of scientists that historians and anthropologists (*and presumably not scientists themselves*) are able to address what it is that scientists do, further suggesting in reference to *in vitro* cell cultures that "it takes an anthropologist in the laboratory to note the strangeness of what has become quickly routinized or banal to its practitioners."[3] Landecker assumes that most scientists are not aware of or changed by the daily activities found in a lab. I agree that many scientists may not be sufficiently reflexive. But this is not the case for all scientists. In fact, this chapter is dedicated to exploring a pause that was generated precisely by such a strangeness. The pattern of growth, division, and purposely inflicted death that took place every three to four days in the *in vitro* cell line of neurons that I worked with created a distinct temporal cyclicity, which although was unlikely to occur anywhere *in vivo*, nevertheless became part of my own basal rhythms. Just as I was responsible for designing experiments using these cells, the patterns of growth and multiplication of these neurons, and their cyclic hormone secretions, regulated my life for years.

On that particular day when I paused to open the incubator door, this rhythm was interrupted as the weight of several entanglements I could not continue to ignore came to bear on me, on the stainless-steel handle of that incubator door, and on the cell line of gonadotropin releasing hormone (GnRH) neurons known as GT1-7 cells. These entanglements included acknowledging accountability for the human practice of developing cancer in mice and killing animals and cells for research purposes. Obtained from the brains of transgenic mice that had been created through the technique of targeted tumorigenesis, these particular *in vitro* neurons were genetically designed to express an oncogene that led to the development of hypothalamic tumors.[4] Before I even came to work with these neurons, there was already an entanglement with a molecular biology technique that involved the noninnocent human-mediated infection of polyomavirus isolated from monkey kidney cells to produce SV40 T-antigen, which caused malignant transformation of the infected cells. There were further entanglements of a reproductive justice movement and an antagonist and partial agonist of the progesterone receptor RU486 that was behind the pause I experienced that day.[5] While leading many feminists to march in the streets for reproductive justice, the compound RU486 led me straight into a molecular biology and reproductive neuroendocrinology lab with the desire to learn more about the molecular mechanisms involved in the regulation of reproduction. Once there, I found myself intimately entwined in a molecular relationship with this *in vitro* cell line of neurons.

As significant as that pause was for me that day, and despite the fact that I knew I had to confront this challenging but generative ontological quandary, I did not have the language or skill set at the time to articulate my question about the lives and responses of *in vitro* neurons in a way that would be recognizable to my colleagues in the sciences. When I think about my interest in the response of those *in vitro* neurons in the lab, I realize now that even back then I was on my way to becoming an interdisciplinary scholar. The examples of interdisciplinary work available to me at the time were not particularly amenable or easily translatable to the inquiry I had generated as a feminist scientist. A new era of interdisciplinary exchange between the humanities and the sciences has helped to relieve some of this unintelligibility.

Interdisciplinary Incubations

These are indeed exciting times for interdisciplinary scholarship. For me, the pauses and causes for reflection encountered by trying to bring together the sciences and humanities can be incredibly complicated but also generative. Over the past decade, I have found a space within the field of women's studies to carry that pause I experienced in a molecular biology and reproductive neuroendocrinology lab along with me and to begin articulating that moment as a meaningful one, worth further reflection. As an invested onlooker and participant in feminist STS, I have witnessed significant paradigm shifts not only within this subfield in regard to its relationship with the sciences, but more generally also in the broader discipline of women's studies itself. These paradigm shifts may be connected to the intellectual fallout from the science wars that took place in the mid-1990s as well as the ever-increasing uneven distribution of funding between the sciences and the humanities. Regardless, it is safe to say that the discipline of women's studies (along with many other humanities-based disciplines) is currently undergoing significant paradigm shifts to reorient itself in relation to the "hard" and natural sciences. The subfield of feminist STS serves as an example of the magnitude of these reorientations, as scholars wrestle not only with the fast-paced development of new biotechnologies but also with the impacts of recent ontological, posthumanist, and material turns in women's studies and the humanities at large.

The critique of poststructuralism's influence on feminist theory and its apparent inability to deal with matter itself has brought forward calls for developing new types of engagements with biology—namely, through scholarship in material feminisms and feminist new materialisms.[6] These calls have brought with them an era of enlivened regard for the sciences. Moving from in-depth critiques of gendered language and the use of gendered paradigms in science, to mining scientific research and data in efforts to move feminist theory forward, there has undoubtedly been a significant shift in the tone with which some women's studies scholars (particularly those who are not trained in the sciences) now voice their interest in the sciences. Having placed questions that are central to the humanities in exchange with research in the natural sciences, this era of interdisciplinarity has made the question that I posed that day in the lab—What would these neurons think?—while holding on tight to that incubator door, somewhat more legible. However, this increased exchange between the humanities and the sciences has also precipitated two major challenges for feminist STS scholars: first, being able to acknowledge life and the living at the level of the *in vitro*, the molecular, and even the inorganic; and second, being able to respond to and deepen our human entanglements with these very lives by paying attention to questions of context and calls for social justice.

Just as an earlier wave of interdisciplinary work in the humanities forced us to examine the question of what it means to be human through multiple and inevitably intersecting frames of sex, gender, race, class, sexuality, ability, and more, as a result of new exchanges with the sciences, the first challenge that the next generation of feminist scholars must face is to trouble the central premise of this very question. The current generation of feminist STS scholars is working hard to learn about the natural sciences, but not simply to find ways to understand the human condition alone. Rather, new alliances between the natural sciences and such fields as women's studies are in fact working to decenter the question of the human within the humanities. Sustained entanglements with animal behavior research, evolutionary biology, molecular genetics, and more have complicated our understandings of exactly what gets to count as a "life" and which lives are included in our concerns regarding "expressive life."[7]

The growth of posthumanist and animal studies is an indication of this paradigm shift when it comes to thinking about the ontological contours of "life." In the field of women's studies, for example, distinctions between the human and nonhuman and the living and nonliving have been troubled by Donna Haraway's idea of naturecultures, Karen Barad's theory of agential realism, Jane Bennett's notion of vibrant matter, and more recently, Mel Y. Chen's concept of animacy, to name but a few.[8] From these recent theoretical moorings, a question that seems important to address, when considering the future of feminist STS in this new era of interdisciplinary work, is not only whether we can continue to ask what it means to be human, but whether our theoretical frameworks and methodologies are prepared to support the question of what it means to be a life—nonhuman, inorganic, and otherwise. This chapter tries to better appreciate what this notion of life and life in the lab can mean by turning to research in synthetic biology. A field borne through molecular biology practices, synthetic biology has produced both *in vitro* and synthetic lives, made of assemblages of both organic and inorganic matter. These lives, known as minimal genome organisms or minimal cells, move across taxonomical thresholds. Interestingly, Michel Foucault wrote in *The Order of Things*, "Up to the end of the eighteenth century, in fact, life does not exist: only living beings. . . . As for life and the threshold it establishes, these can be made to slide from one end of the scale to the other, according to the criteria one adopts."[9]

It is evident that the line we draw between the living and nonliving can be made to slide according to the criteria we adopt. Similarly, Deleuze and Guattari have described life as an intensive and powerful life that is not organized or necessarily found within an organ or an organism. In the final plateau of *A Thousand Plateaus*, Deleuze and Guattari dedicate their focus to the topic of space, discussing several modes of smooth (read as nomadic) and striated (read as sedentary) space, including technological, musical, maritime, mathematical, physical, and aesthetic space. Their discussion of aesthetic space emphasizes the mixture of smooth and striated space at all times, noting the lack of creativity that accompanies those lines that have been drawn to confine life within the organic alone. "If everything is alive," they wrote, "it is not because everything is organic or organized but, on the contrary, because the organism is a diversion of life. In short, the life in question is inorganic, germinal, and intensive, a powerful life without organs, a Body that is all the more alive for having no organs, everything that passes *between* organisms."[10]

I turn to the idea of the body without organs below, but here I want to emphasize that the technologies and practices of synthetic biology have most certainly redrawn lines and disrupted what we have come to consider as the thresholds of life. This disruption may be jarring to some but, similar to the pause that interrupted me while splitting *in vitro* neuronal cells in the lab that day, I believe it is important to reflect on this disruption and consider the possibility that these synthetic lives may not only be expressive but may also be reaching back out toward the surrounding world. The question is, Will we know how to respond?

The second challenge for feminist STS scholars arising from this interdisciplinary exchange is a direct result of the earlier challenge. As all eyes turn to the nonhuman and to molecular and subatomic matter, we must remain aware of the costs of building theoretical interventions apart from their human social and political implications and entanglements. Our ideas of the social and political can expand so as to include or even focus upon the nonhuman, but as we shift our central questions in the humanities, we must keep in mind the broader contexts and repercussions of this work. Thinking with the nonhuman or even the inorganic is an ethical project, but it does not mean, as the chapter epigraphs suggest, that we must leave humans behind or, for that matter, our notions of social justice behind. What is key here is *not to stop* theorizing once we have initiated our ontological, posthumanist, and material turns. We must keep theorizing our way through until we can connect these new insights to our role and contributions as humans within these turns. In addition, our ideas of social justice must apply to all forms of life—from humans to nonhumans, from the organic to the inorganic. Haraway refers to this in her work as "multispecies ecojustice."[11]

In her work on Darwin, feminism, and sexual difference, Elizabeth Grosz asks us to consider the following: "How does biology—the structure and organization of living systems—facilitate and make possible cultural existence and social change?"[12] As a biologist, I am on board with the idea that biology can be used to initiate social change and even work toward social justice. In fact, an antagonist and partial agonist of the progesterone receptor are precisely what made me march straight into a molecular biology lab in the first place. I am committed to what feminists can come to know not just by collaborating with the sciences but also by collaborating with molecules. I am invested in the futures we can begin to imagine by turning to the practices of the biological sciences and to the capacities of biological matters. However, I also think that much about what we come to know and the future that we want to see depends on the specificity of which "social change" we are talking about, and the approaches we actually use to get there. Envisioning new ontological and ethical frameworks is difficult work as it is, but biophilosophies of becoming require that we figure out how to apply these frameworks and live with these difficulties.

Vicky Kirby has made a compelling case that bacterial cells write and that "it is in 'the nature of Nature' to write, to read and to model."[13] Her intervention is crucial in terms of the first challenge posed to interdisciplinary scholars in the humanities and sciences—namely, what it means to be nonhuman. For feminist, postcolonial, and decolonial STS scholars, who are committed to thinking about social justice and want to make connections between the humanities and sciences, such an ontological intervention must be expanded to meet the challenge of thinking about contextual accountability. In her work on Niels Bohr, Albert Einstein, Werner Heisenberg, and quantum physics, Karen Barad draws from Jacques Derrida's idea of "justice-to-come" to discuss entanglements and the behavior of atoms. She states:

> The past is never closed, never finished once and for all, but there is no taking it back, setting time aright, putting the world back on its axis. . . . The trace of all reconfigurings are written into the enfolded materialisations of what was/ is/ to-come. Time can't be fixed. To address the past

> (and future), to speak with ghosts, is not to entertain or reconstruct some narrative of the way it was, but to respond, to be responsible, to take responsibility for that which we inherit (from the past and the future), for the entangled relationalities of inheritance that "we" are. . . . Only in this ongoing responsibility to the entangled other, without dismissal (without "enough already!"), is there the possibility of justice-to-come.[14]

Barad emphasizes the importance of recalling the past and thinking about the responsibilities we as humans hold while trying to think differently about materiality.

Gill Jagger has recently argued that compared to Barad's theory of agential realism, Grosz's and Kirby's turns to nature fall short of providing a useful way of rethinking materiality. "Thus, if the aim of the new materialism is to provide a way of rethinking the interimplication of culture and nature," Jagger writes, "moving away from the negation of one in the determination of the other, difficulties remain in both Kirby's and Grosz' accounts. This is not the case, however, with Barad's account of the intra-action of nature and culture in the material-discursive relation: it involves a process of mutual articulation that is a matter of interimplication."[15] In my understanding of their work, Grosz and Kirby encourage us to think differently with the sciences to imagine biology and "nature" as providing the grounds for social change. In their own ways, Grosz, Kirby, and Barad each encourage us to find new ways to think about the relationships between human, nonhuman, organic, and inorganic lives through closer analyses of science, biology, physics, and matter. Perhaps because of Barad's training as a scientist, however, I would agree that her work may resonate more with the feminist scientist who is concerned with questions of interimplication. For feminists who are working directly in scientific disciplines, recalling the past is a reflexive practice that requires learning how to think about the context of a biological event as deeply and broadly as possible.

Roots and Shoots: Approaches to Life and Context

As part of the basic structure of many species of grass, horizontal stems known as rhizomes (stems that grow below ground) and stolons (stems that grow above the ground) can form "nodes," which in turn can give rise to both "roots and shoots." These new roots and shoots can develop new "daughter" plants.[16] Similarly, molar and molecular politics can come together and form new projects in feminist STS. I have already started recounting my inquiries into the lives and responses of nonhumans by discussing my encounter with an *in vitro* cell line of neurons, developed through targeted tumorigenesis, and used for molecular biology research. This chapter brings into the mix other *in vitro* bio-actants such as minimal genomes, and bacterial and yeast cells that are referred to as surrogates in synthetic biology research. Yet before focusing on the intimacies of *in vitro* life, some may have more pressing concerns whether as humans we should be tampering with genes and organisms, or "playing god" at all.

For some of us, the more acceptable and familiar place to begin this interdisciplinary analysis might be to interrogate the processes that have led to the recombination of genes, the creation of transgenic animals, and the synthesis of new organisms in the first place. Some of us may also question the tenets of molecular biology itself and the validity of a science that places such authority and focus on DNA. Others might be more than wary of the pervading reductionist logic that lies behind molecular biology as a whole, which has resulted in the field of synthetic biology and forwarded a completely mechanistic view of life. Alternatively, some of us may be raising traditional bioethical concerns related to agency, choice, and the safety of conducting recombinant DNA, transgenic, and synthetic biology experiments. Of course, these concerns are valid and require much deliberation. Although these concerns are crucial, they also follow already well-established lines of inquiry between feminism and the biological sciences. They are molar in their approach not only because they represent tried-and-true modes of inquiry but also because in many cases they eventually return us to questions of human subjectivity, identity, and representation.

Once again, I want to be clear that in feminist STS, it is necessary for scholars to continue their interdisciplinary work through such molar approaches and lines of inquiry. However, it is also necessary for some feminist STS scholars to take more molecular approaches to their inquiries at the intersections of feminism and molecular biology. These molecular approaches may not begin by assessing the "appropriateness" of a science such as synthetic biology, or whether it is ethically "correct" to create transgenic cell lines or animals, or use bacteria as surrogates to create minimal synthetic bacterial cells. In this case, a molecular approach might be about spending more time learning about the intricate practices of synthetic biology in order to look for places where feminist and scientific questions of life, matter, context, justice, and ethics may be placed together.[17] A molecular approach may also involve suspending (even if just for a moment) our capital "E" ethical judgments regarding whether synthetic organisms should exist at all. It is perfectly sound to ask the question, How did we arrive at this point? However, if we follow a molecular approach, the question might become, Now that we are here, what is our relationship with the synthetic lives that already live among us?

My tendency to turn toward molecular feminisms obviously stems from the pause I experienced in the lab while working with an *in vitro* cell line. However, it also reflects that long-standing interest I have had in thinking about biological matters in the lab through biophilosophies of becoming and reaching toward these matters through microphysiologies of desire. So far, I have focused on the qualities of changefulness, nonhuman becomings, kinship, and hylozoism in our encounters with nonhuman others such as grass and bacteria. In this last chapter I think about ways to encounter synthetic lives that are already our kin by highlighting the qualities of univocity and immanence. Although I barely begin to scratch the surface, I pursue this encounter in two ways: first, by posing the question of what constitutes life and the living in this era of synthetic biology; and second, by finding a way to consider the human entanglements that contribute to the contexts in which these lives are lived.

In *After Life*, Eugene Thacker has traced the ontology of life through a history of Western philosophy. He states:

> "Life" is a troubling and contradictory concept. . . . Today, in an era of biopolitics, it seems that life is everywhere at stake, and yet it is nowhere the same. The question of how and whether to value life is at the core of contemporary debates over bare life and the state of exception. At another level, in our scientific worldview, it seems that life is claimed of everything, and yet life in itself is nothing. While biologists continue to debate whether or not a virus is living, the advances in genetic engineering and artificial life have, in different ways, deconstructed the idea that life is exclusively natural or biological.[18]

Thacker suggests that there are three major modes through which philosophical engagements with the question of life are organized today. They are the affective-phenomenological, the biopolitical, and the political-theological.[19] It is within the affective-phenomenological mode that we find those approaches that relate to a biophilosophy of becoming and turn to the "immanently dynamic, self-organizing, and germinal qualities" of life.[20] Life in this sense is treated as an event, a proliferative one at that, bringing with it the capability of generating difference.

Thacker argues that our habits of thinking about life in a hierarchical fashion, beginning with biological elements and building layers of behavior, culture, and politics up onto this scaffold, is a direct result of Aristotle's philosophy of life. Starting with the philosophical works of Aristotle, Thacker turns to concepts of life that have attempted to work against this stratification. Although his project ultimately points to some inherent contradictions that are constitutive of the various concepts of life, Thacker provides a rich analysis of the importance of such concepts as univocity and immanence to understanding life in the biophilosophy of Deleuze. We know that for Deleuze, univocity is understood as a univocity of difference. The concept of univocity is what drives Deleuze's ontology and is crucial to the idea that life can exist as a multiplicity within an ontologically single field. As Thacker explains:

> In *Difference and Repetition*, Deleuze takes up the concept of univocity in a way that places it at the center of his ontology of difference. For Deleuze, traditional ontology is predicated on the concept of identity (vs. difference), of the One (vs. the Many), of Being (vs. becoming), and so on. That-which-differs can be regarded only as in some way falling away from, or dependent upon, that which does not differ, or that which is whole, Ideal, One. As Deleuze states at the outset, his aim is to think the concept of difference not as secondary or derivative, but in some way as primary to our thinking about that-which-differs as well as to the processes of differentiating and creating differences.[21]

This framing of life through univocity, where univocity plays a central role in an ontology marked by difference, is the molecular approach to life and the living that is taken up in this chapter. Synthetic life therefore can be understood as an event, with the capacity for generating difference. I am particularly interested in the lives that have been created via human, bacteria, DNA, protein, and technological interactions, ones that notoriously zigzag across taxonomical lines (between the organic and inorganic, living and nonliving), and ones that are expressive. Although they are synthetic, there is much to learn from minimal genome organisms, if we are able to work with them through this concept of univocity and if we can orient our curiosities toward them upon an immanent plane.

In this synthetic era of biology, we are being pressed to reconsider our onto-ethical orientations toward lives that are beyond being merely recombinant or transgenic. Indeed, "oncomouse" has now become an elder at the table.[22] Minimal genome organisms, designed and produced synthetically in a lab, are taking us out of our previous comfort zones, demanding that we revisit and further expand our notions of kinship. The concept of immanence may help to bring us to this newly reconfigured table. "For Deleuze," as Thacker explains, "this conjunction of immanence and expression—or really, *of immanence and life*—has three fundamental principles."[23] These principles include the principle of equality, where "immanence is not only the immanence between Creator and creature (a vertical immanence), but the immanence between creature and creature (a horizontal immanence)." It also includes the principle of univocity as discussed earlier, which allows for an immanence that is "at the same time dispersive and inventive, distributive and creative, supernatural and natural." Lastly, immanence can be characterized by the principle of affirmation, incorporating "an ontological affirmation that supports a notion of being as purely superlative, affirmative, and creative" and not one defined negatively through lack.[24]

Several paradigms and practices come together to form the field of synthetic biology. The challenge we face as feminist STS scholars with this field is that while it has pushed us to question the boundaries drawn around life and the living, it has also created a novel synthetic life cycle. This synthetic life cycle first travels through a human-mediated and computer-coded inorganic phase, which begins with digital representations of DNA that are used to place molecular materials into synthetic structures such as the minimal genome. It then moves on to an organic phase where minimal bacterial genomes that have been genetically engineered are introduced into "surrogate" cells (whose "naturally" occurring genome has been removed) to become an organism that transcribes, translates, and produce proteins of interest *in vitro*. Lastly, the life cycle progresses to what might be considered a social phase, whereby a minimal genome organism, that has been synthesized to contain genes of human interest, requires a variety of human and nonhuman systems and resources in order to thrive. During this phase new forms of expression such as technologies can emerge as a result of the synthetic life cycle that both organize and are organized by humans and the environment. The challenge is to treat each phase of this life cycle through the qualities of univocity and immanence, beginning with the inorganic phase.

Deleuze and Guattari speak of inorganic life that is expressive and germinal, that exists as a body without organs (BwO). For many scholars the ideas of inorganic life and a body without organs both represent highly contentious aspects of Deleuze and Guattari's work. The turn to inorganic life has been criticized as an attempt to recuperate some form of neo-vitalism.[25] In addition, there has been much confusion around their concept of the body without organs, often being interpreted as a stance against organs themselves.[26] I address both concerns here briefly. For many, vitalism is a highly fraught and untenable philosophical position. In my opinion, the charge of vitalism in Deleuze and Guattari's work, and perhaps in my own project here of thinking with the lives of bacteria, an *in vitro* neuronal cell line, and minimal genome organisms, represents a failure to recognize the important philosophical project of reframing and reimagining life, biology, and matter.[27] In his work on Deleuzian approaches to thinking about life, John Protevi explains, "Deleuze is a machinic materialist, not a mechanist, and it is only as a reaction to mechanism that classical vitalism makes sense. It is the impoverished sense of matter in mechanism, as chaotic or passive, that creates the temptation to classical vitalism of the 'entelechy' type. . . . What we need to look for in Deleuze's notion of vitalism is the 'life' that encompasses both organisms and 'non-organic life.'"[28]

Similarly, the idea of a body without organs can easily be misread. Indeed, Deleuze and Guattari, much like their position toward trees, on a first read appear to be "anti-organ" or anti-organism. Yet when they make such statements as "we're tired of trees" or "the enemy is the organism," they are in fact referring to their position against an arborescent model of linear and hierarchical thought.[29] In the case of the BwO, they are commenting on how life might be better understood by "situating it within the wider field of forces, intensities, and durations that give rise to it and which do not cease to involve a play between nonorganic and stratified life."[30] Leslie Dema has suggested that by using the idea of non-organic or inorganic life, Deleuze and Guattari are attempting to disrupt our habit of creating taxonomical and terminological breaks and that "their theory of life directly challenges the idea of organic life that we find in contemporary biology."[31]

Although the idea of organic life in contemporary biology has been greatly troubled in recent years by the arrival of synthetic biology, Dema makes another crucial point regarding the philosophical challenge that is presented to us by confronting the idea of inorganic life. She explains that the best way to understand Deleuze and Guattari's idea of inorganic life is through their concept of assemblages. Assemblages are, according to Dema, "not like organs" but instead are "animated by coding and decoding, deterritorializations, and lines of flight." She states, "Assemblages are the symbiotic or sympathetic co-functioning of heterogeneous elements. They are formed through a rapport between partial objects that enter into monstrous couplings, experimental alliances, unnatural participations, and rhizomatic structures."[32] It is certainly fitting to characterize the inorganic life that begins the synthetic life cycle as a monstrous coupling or experimental alliance. With the coming together of digital DNA, humans, DNA synthesizers, Petri dishes, bacteria, and yeast, the idea of the assemblage is useful to contextualize a life produced by synthetic biology. Furthermore, the idea of an assemblage, similar to microphysiologies of desire, provides a way to encounter and extend ourselves toward synthetic lives through the qualities of univocity and immanence, and with a methodology to consider questions of context.

According to Deleuze and Guattari, there are two types of assemblages—namely, machinic assemblages of desire and collective assemblages of enunciation.[33] While collective assemblages of enunciation work at the level of language and the symbolic, Levi Bryant explains that "when Deleuze and Guattari refer to machinic assemblages they are talking about the domain of physical objects, how they interrelate, and how they affect and are affected by one another."[34] Thinking about life in terms of machinic assemblages and material objects that come together to influence each other and connect with each other presents an alternative to thinking about synthetic life in only mechanistic terms such as DNA synthesis and protein production. In *Germinal Lives*, Keith Ansell Pearson has suggested that the process of paying attention to machinic assemblages is a key part of Deleuze and Guattari's strategy for approaching life itself. "A 'machinic' approach, then," he states, "will not treat machines as projections of the human but rather in terms of 'monstrous couplings' involving heterogeneous components that 'evolve' in terms of recurrence and communications. . . . Humans are both component parts of a machine and combine with other forms of organic and nonorganic life to constitute a machine (or, better, machinic assemblage since there exists no isolated and monadic machine)."[35]

Reiterating the sentiment from his chapter epigraph, Ansell Pearson sees the machinic assemblage as a togetherness of organic and inorganic forms, and most important, a togetherness where the human is not left behind. By using the idea of the machinic assemblage, and by aligning ourselves with the qualities of univocity and immanence, a biophilosophy of becoming that draws from Deleuze to "think beyond the human condition" does not need to leave humans behind.[36] Deleuze and Guattari describe the assemblage as a multiplicity, and a machinic assemblage more specifically as having one side that "faces the strata, which doubtless makes it a kind of organism" and another side "facing a body without organs, which is continually dismantling the organism."[37] In other words, an assemblage can orient itself toward both molar and molecular tendencies.

In addition to being easily characterized as its own monstrous coupling, throughout its lifecycle a synthetic life demands a great deal of support from human, machine, and environmental resources. To bridge concerns over what constitutes life and living in the synthetic age of biology, with concerns over context and the role that we as humans share in sustaining these life cycles, I suggest we see ourselves and these synthetic lives as part of a machinic assemblage. One side of this machinic assemblage faces toward molar tendencies of stratification, and the other toward molecular tendencies of dis-organ-ization. While discussing Deleuzian approaches to the question of life, Protevi explains: "For Deleuze and Guattari, 'life' has a double sense, reflecting both stratification and destratification. It means both 'organisms' as a certain set of stratified beings and also the creativity of complex systems, their capacity to produce new emergent properties, new behavior patterns, by destratifying and deterritorializing."[38]

Given that so much of the emphasis has been placed on the molecular within this text, the remainder of this chapter addresses the second challenge of finding ways to consider questions of context by reflecting on those elements of the machinic assemblage that face the molar or the side of stratification. As feminist STS scholars, we may not personally have a hand in creating new lives in this synthetic era of biology. However, we can begin to see ourselves as part of a machinic assemblage that includes these synthetic lives. If indeed we do begin to see ourselves as part of such an assemblage, there is a possibility to think differently about our role within that assemblage. We need to learn how to use both molecular and molar approaches in order to live and respond to those synthetic lives that are already here among us. This is perhaps one way to become accountable for our part within an entanglement or, as Barad says, aware of our responsibilities for "that which we [have come to] inherit."[39]

Inorganic Stratum: The Central Dogma and Its Implicit Forms

In the foreword to *The Order of Things*, Foucault states: "It is not always easy to determine what has caused a specific change in a science. What made such a discovery possible? Why did this new concept appear? Where did this or that theory come from? Questions like these are often highly embarrassing because there are no definite methodological principles on which to base such an analysis. The embarrassment is much greater in the case of those general changes that alter a science as a whole."[40]

Here Foucault suggests the difficulty in tracing the factors involved in a specific change in a science, or in other terms perhaps, the birth of a new paradigm. This is true, particularly in the case of tracing how scientists have come to think about life and what constitutes the attributes of the living. In 1958, however, an important event occurred, a kind of big bang one might say, that altered the future of molecular biology and genetics.[41] This event was the formulation of the central dogma. As embarrassingly simple the following tracing of the central dogma may be, I turn to it here because of its resemblance to expressions that can be found in inorganic strata. Protevi explains that "in the inorganic strata, expression is the molarization of molecular content that is, the carrying forth to the macroscopic scale of the 'implicit forms' of molecular interactions."[42] While the central dogma cannot be easily determined as the one and only implicit form that changed how biologists went from thinking about molecules to macroscopic organisms, it is a significant event worth remembering for its role in shifting or "molarizing" how scientists have come to think about intensive or expressive aspects of life.

The central dogma refers to the process of the unidirectional and sequential flow of genetic information originating from DNA, moving to RNA, and then from RNA to protein.[43] DNA and RNA are both biopolymers that are made of nucleic acids and are comprised of the four nucleotides adenosine, guanine, cytosine, and thymine in the case of DNA, and adenosine, guanine, cytosine, and uracil in the case of RNA. The central dogma tells us that both DNA and RNA provide the code for protein synthesis, which occurs through a two-step process. The first step is referred to as transcription, whereby DNA serves as a template for the production of single strands of messenger RNA. The idea is that the information that is coded on the DNA template is transcribed or, similar to how the term is used in computer science, is transferred from one recording system to another. This transfer of DNA code is mediated by the enzyme RNA polymerase, which works to produce a new kind of information or code—one that is in the form of a complementary and antiparallel RNA sequence. For example, an antisense strand of DNA such as 3'ATGACGGA5' is transcribed into the sense mRNA strand 5'UACUGCCU3'. This newly synthesized RNA molecule, however, is simply another messenger or a go-between, destined only to deliver the command required for gathering amino acids in the final event of protein synthesis.

This next step of the mechanism of moving from code in messenger RNA to an ultimate protein destination is referred to in the field of molecular biology and genetics as the process of translation. Translation is an important in-between process that proceeds in four phases, including activation, initiation, elongation, and termination. During these four phases of translation, a series of three nucleotide base pairs come together to create what is called a codon. Each codon is then decoded by a ribosome, and with the help of transfer RNA (tRNA) molecules, a chain of amino acids come together to form a protein. In synthetic biology, scientists make use of the metaphor of code to write or to program this code in a specific way and thereby have a hand in directing protein biosynthesis. Interestingly, if we recall Judith Butler's concern of "What of life exceeds the model?" noted in an earlier chapter and voiced during an interview with Vicky Kirby, we can see that the metaphors and models of coding, transcription, and translation, which have been used to explain and carry out the central dogma, have long become the ontology of molecular biology and life itself—and quite productively, I might add.[44]

The central dogma in molecular biology was created and so named by the scientist Francis Crick, biophysicist and codiscoverer of DNA's molecular structure. Prior to his collaboration with James Watson, Crick had been trained as a physicist and was working on the X-ray crystallography of proteins. However, starting in the 1940s, there was immense interest and growing excitement in the field of protein biochemistry, in great part due to the work of Linus Pauling, who had also been trained as a physicist. In 1945, Pauling submitted a grant to the Rockefeller Foundation to launch a research program that was to become the field now known as molecular biology.[45] Pauling was also responsible for popularizing the application of quantum physics into chemistry, in addition to developing the practice of 3-D molecular modeling. Since Pauling's day, the practice of protein modeling has moved from plastic balls and wooden sticks to highly complex computer modeling. I mention Pauling's and Crick's common background in physics and their shared interests in the physical and mechanistic aspects of protein chemistry and the structural modeling of proteins, because of what I see as an interconnected set of events that sheds light on a dominant paradigm that currently guides the field of synthetic biology and its purchase on life and the living.

When asked why he named the process of information transfer, from DNA to RNA to protein, the "central dogma," Crick apparently admitted to his mistake and laughed at his misunderstanding of the meaning of the word "dogma." In his autobiography, *What Mad Pursuit: A Personal View of Scientific Discovery*, Crick stated:

> I called this idea the central dogma, for two reasons, I suspect. I had already used the obvious word hypothesis in the sequence hypothesis, and in addition I wanted to suggest that this new assumption was more central and more powerful. . . . As it turned out, the use of the word dogma caused almost more trouble than it was worth. . . . Many years

> later Jacques Monod pointed out to me that I did not appear to understand the correct use of the word dogma, which is a belief that cannot be doubted. I did apprehend this in a vague sort of way but since I thought that all religious beliefs were without foundation, I used the word the way I myself thought about it, not as most of the world does, and simply applied it to a grand hypothesis that, however plausible, had little direct experimental support.[46]

Crick's central dogma continues to serve as a particularly powerful structural and functional paradigm. Even with the now accepted phenomenon of epigenetics, the linearity and simplicity of the central dogma serves as a cornerstone for understanding the organization of organic life and the emergence of proteins through the processes of transcription and translation. Crick's translation of the term dogma, or shall I say "mistranslation" of the term, has had profound ontological, epistemological, methodological, and ethical impacts on how we orient ourselves while dealing with DNA, cells, and organisms in the lab. It has had a profound influence on how we think about life itself in biological terms and how molecular content is "carrying forth to the macroscopic scale."[47] Molecular biologists have relied on the central dogma to develop recombinant DNA technologies to design and bring forward new lives such as transgenic organisms. The central dogma has served, if not as a religious belief, then as a highly revered principle for many scientists. It can be argued that the paradigm of the central dogma has provided the intellectual anchor for a number of additional scientific enterprises on a global scale, including the justification for spending billions of US taxpayer dollars to fund the Human Genome Project. It has also provided the scientific authority needed to continually drive social arguments based on genetic determinism, as is evident in the rise of a new eugenics.[48]

That the term "dogma" is generally understood as that which is authoritative and not to be disputed, but is simultaneously a belief that originates without reason or evidence, is not the meaning that Crick understood in his naming of a particularly important sequence of molecular events. I argue, however, that this is exactly how the central dogma has operated and continues to operate in molecular biology and, most effectively, in the field of synthetic biology. The idea that a unidirectional, linear, and hierarchical deployment of molecules inside an organism is required for the structure and formation of life lays the intellectual foundation for the field of synthetic biology. Synthetic biologists, who apply engineering principles to the design and creation of new life forms, were raised on molecular biology's central dogma. Since the early 1970s and the advent of recombinant DNA technologies, scientists have been working on altering life forms. They have been guided by the central dogma but have also been taking advantage of the fact that DNA can be cut or digested with restriction enzymes, altered through the insertion of a foreign or synthetically produced piece of DNA, and then ligated back together. For instance, for decades now, molecular biologists have designed and used transgenic or knockout animals to understand the biological basis of human diseases. These animals have been designed to contain mutations in a specific gene, contain a completely "foreign" gene, or have a gene completely deleted in order to study a gene's function and correlation to human disease.

However, many molecular biologists have grown weary of the arduous hit-and-miss techniques of recombinant DNA technologies. These scientists are turning to the new tools of synthetic biology to study the material processes of biology. For synthetic biologists who see themselves as bioengineers, the beauty and simplicity of the central dogma lies in the fact that molecules such as DNA, and in turn molecular life, can exist in an inorganic form as a language or computer code. Instead of having a binary code of 1's and 0's used in computer processing, the main biological components of life are thought to be comprised of the four letters A, G, C, and T. The BioBricks Foundation, for instance, literally stores DNA as a code, and thousands of inorganic gene cassettes can be transferred onto a computer hard drive, in the form of magnets and megabytes. The foundation's goals are to make DNA (as inorganic and digital code) accessible to everyone and, as a result, create a better world through biology. The foundation's website explains:

> Biology is everywhere. And matters to everyone. It affects our food, medicines, homes, and environment. Yet people are not working well together as partners with biology. BioBricks Foundation believes in a future where there is a free-to-use language for programming life that benefits everyone. A future where people around the world communicate and collaborate to create local biological solutions to meet global

needs. When people are inspired to work in partnership with biology this future is possible. When people have the tools and infrastructure to work with one another, we can meet global needs for food, medicines, shelter, clean water and air.[49]

In an effort to create such tools, there are three main ways of applying engineering principles to the material reconstruction of DNA—namely, the bottom-up, the top-down, and the pathways approaches.[50] The research of a few prominent synthetic biologists, whose work represents diverse interests in the field, is discussed briefly here to reveal the logic behind each of these three approaches.

The bottom-up or "parts-based" approach to creating synthetic life can be characterized by the work of Drew Endy, a civil and biochemical engineer, previously at MIT and currently at Stanford University. Endy, who is pushing for the creation of an open-source platform for genetic biotechnology, is the founder and president of the not-for-profit BioBricks Foundation. His bottom-up approach is based on forward engineering or the idea that DNA can be broken down into separate entities or cartridges that can then be used to deliberately assemble a specifically fashioned or desired biological product. Endy states:

> Consider that most early discoveries of genetically encoded functions depended on analysis of the linkage between natural or randomly generated mutations and phenotypes, a powerful approach akin to blindly smashing many cars with a hammer and then determining which broken parts matter by attempting to drive each machine. Over the past 30 years, the invention and development of DNA sequencing technology have provided a complementary approach for discovering genetic functions. . . . However, two additional approaches are needed to confirm and exhaustively identify all functions encoded by a natural DNA sequence. Specific DNA sequences thought to affect phenotypes must be purposely changed and the expected effect confirmed. Also, seemingly irrelevant DNA sequences must be removed, disrupted, or otherwise modified and shown to be unnecessary. . . . Going forward, the ability to implement many simultaneous and directed changes to natural DNA sequences and to build and test synthetic systems will give researchers a powerful new "hammer" for constructing how life works.[51]

Alternatively, the top-down approach is led by entrepreneur and geneticist J. Craig Venter, whose work is discussed in greater detail below. In contrast to the bottom-up approach, Venter's top-down approach can be summarized as starting with full genomes and then scaling them down to a minimal size, such as in the case of the minimal genome used to create the first minimal synthetic bacterial cell.[52] This approach has been described as being modeled upon a "plug-and-play" set of functions.[53] The last approach, roughly named the pathways approach, is illustrated by the work of Jay Keasling, professor of chemical engineering and bioengineering at UC Berkeley and associate director of the Lawrence Berkeley National Laboratory. Keasling's work, utilizes a pathways or problem-driven approach and opts for the use of any and all engineering approaches that make the modification of DNA more practical and cost-effective.[54] Regardless of their technical differences, what we are witnessing in this stage of the synthetic life cycle and from this particular orientation of the machinic assemblage is, as Protevi has called it, the "molarization of molecular content."[55] Each of these approaches falls in line with the workings of the inorganic stratum. We can see here exactly how far an implicit form, by way of the central dogma, has shaped and produced our knowledge regarding how molecules interact and how molecular structures can come together to form synthetic macromolecules. This implicit form has given birth to synthetic life.

Organic Stratum: Minimal Lives Respond to Problems

In *The Politics of Life Itself*, Nikolas Rose posits the politics of "life itself" as the vital politics of our time. He is concerned with the growing capacities to "control, manage, engineer, reshape, and modulate" the vital capacities of human beings as living creatures.[56] Defining the idea of vital politics he wishes to put forward, Rose describes what he calls a major shift in biopolitics today compared to the first half of the twentieth century. He suggests that recent developments in molecular biology have led to the phenomena of a "molecular vision of life."[57]

Rose is concerned with tracing an emergent form of life and biopolitics that foreground the human, but I argue that synthetic biology requires us to trace a different concept of "life itself" as it relates to the emergent capacities of nonhuman minimal genome organisms. The concept of "life itself" that must be applied here relates not only to the organic bacterial "surrogates" that are part of the machinic assemblage but also to inorganic life and minimal genomes that come together to form synthetic life. If any life can be said to have gone through the phenomena of a "molecular vision of life," bacteria would have to be at the very top of this list.[58] The step-change in life that Rose argues we as humans have experienced is simply a change in scale, from whole organisms to the molecular parts of whole organisms. This change in scale is shadowed by the step-change that has occurred at the level of bacterial microorganisms. The step-change in life that I am referring to has occurred at the level of type, not scale. It is a step-change at the level of type because with the advent of synthetic biology, the definition of "life itself" is being shifted from an organic life contained within an organism, to an inorganic life that begins without organs, can be dis-organ-ized, and is comprised of code. Therefore, in addition to attending to what Rose sees as the extended reach of contemporary biopolitics and a "molecular vision of life itself," this step-change at the level of type forces us to appreciate the *expressive life of a molecule itself*, as synthetic biologists have already done.

If we approach the expressive life of a molecule as a machinic assemblage, we can begin to align our curiosities along the qualities of univocity and immanence. Returning to the work of Deleuze and Guattari, we can think about the second phase of the synthetic life cycle as facing the organic stratum, where "expression becomes autonomous in the linear genetic code, which results in greater deterritorialization (greater behavioral flexibility) of organisms."[59] As Sara Dawn Eimer has described it, the organic stratum is where "the form of a line of DNA, *itself molecular* . . . operates upon other molecules to produce the 'molar' entity, the organism."[60] In synthetic biology, entire genomes have been created through the top-down method of molecular manipulation.[61] These genomes are minimal in the sense that they have been designed to contain the minimum number of genes required for bacterial cell viability and growth.[62]

Minimal genomes may contain the minimal number of genes required for a cell to grow and replicate, but they still require other nonchromosomal elements for these genes to be transcribed and translated into proteins. The process of inserting a minimal genome into a "host" or "surrogate" bacterial or yeast cell whose own genome has been removed is referred to as "genome transplantation."[63] For example, the newly arrived "minimal synthetic bacterial cell" created by the J. Craig Venter Institute is comprised of a synthetic minimal genome that is based on the genome of the bacteria *Mycoplasma mycoides* but has been transplanted into the bacteria *Mycoplasma capricolum* whose genome, in turn, has been removed.[64] Through the process of genome transplantation, we see the coming together of a genetically engineered minimal genome or line of DNA and a surrogate cell. This allows the machinic assemblage to shift its orientation from the inorganic stratum to the organic stratum, moving it from the form of a line of DNA to a molar entity that can be identified as an organism. The minimal genome becomes a minimal genome organism thanks to its surrogate.

It is precisely due to their capabilities of replicating DNA, transcribing DNA, and translating RNA into proteins that bacteria and yeast have long been perceived as potential surrogates or machines in molecular and synthetic biology. For instance, in 2012 a team of synthetic biologists based out of the University of Nottingham announced their intention to create an operating system for new cellular life forms. Their project—named Towards a Universal Biological-Cell Operating System (or AUdACiOuS for short)—was supported through a $1.58 million grant awarded by the Engineering and Physical Sciences Research Council in the UK. This project treated *E. coli* as an "information processing machine."[65] It was aimed at creating a line of bacterial cells that contained the minimal requirement of components to stay alive, but that could easily be programmed to execute specific functions through protein biosynthesis. Natalio Krasnogor, the primary scientist on this project, summarized the research as follows:

> A living cell, e.g. a bacterium, is an information processing machine. It is composed of a series of sub-systems that work in concert by sensing external stimuli, assessing its own internal states and making decisions through a network of complex and interlinked biological regulatory networks (BRN) motifs that act as the bacterium neural network. A bacterium's decision making processes often result in a variety of outputs, e.g. the creation of more cells, chemotaxis, bio-film formation, etc. It was recently shown that cells not only react to their environment but that they can even predict environmental changes. The emerging discipline of Synthetic Biology (SB) considers the cell to be a machine

that can be built—from parts—in a manner similar to, e.g., electronic circuits, airplanes, etc. SB has sought to co-opt cells for nano-computation and nano-manufacturing purposes. During this leadership fellowship programme of research I will aim at making *E. coli* bacteria much more easily to program and hence harness for useful purposes. In order to achieve this, I plan to use the tools, methodologies and resources that computer science created for writing computer programs and find ways of making them useful in the microbiology laboratory.[66]

Sophia Roosth has also discussed the capabilities of microorganisms—namely, the ability of yeast to scream.[67] Although Roosth does not refer specifically to synthetic yeast, she argues that scientists who work with the technique of sonocytology on yeast species such as *Saccharomyces cerevisiae* make a distinction between yeast cellular signaling and so-called baseline or background noise by approaching yeast cells as "subjects capable of speaking to their conditions."[68] Despite the fact that these scientists treat bacteria as machines, they also note the wide range of capabilities that bacteria have, including the capacity to react or respond to their environment. In the organic stratum we start seeing self-organization and that "life responds to problems by experimenting with different kinds of solutions."[69] As it turns out, synthetic organisms show indications of having behavioral flexibility and problem-solving skills.[70] The problem is that they need to learn how to cooperate with one another.

Maitreya Dunham has raised a crucial aspect of this phase of the synthetic life cycle in her commentary "Synthetic Ecology: A Model system for Cooperation." She writes:

Synthetic biology offers the promise of a better understanding of biological systems through constructing them. Unlike naturally occurring biological systems, which are generally complicated by multiple variables and difficult to isolate components, synthetic systems can be simplified to allow for experiments that would be too difficult to interpret if done in their full natural context. Up to now, synthetic biologists have primarily focused on gene circuits . . . [learning] more about the rules of gene expression and regulation, including fundamental issues regarding noise, timing, and signal fidelity. In this issue of PNAS, *Shou et al.* demonstrate an example of a new direction for synthetic biology, what

> might be called synthetic ecology. Rather than using gene modules as building blocks, they mix cell populations to construct a synthetic simple obligatory cooperative ecology.[71]

Our understanding of biology has come to this. Lives that exist as "naturally" occurring systems are far too complicated. However, even though these systems can be simplified through synthetic biology, once they are created, they need to be able to respond to the problem of how to cooperate with each other in order to live in a broader ecology.

Scientists, and particularly molecular biologists, have used the linearity of the central dogma alongside the principles of reductionism to gather more details about the natural world. With synthetic biology, however, we are witnessing something new. What we have here is an ontological premise based on reductionism (DNA, RNA, and proteins) and linearity (transcription and translation) that has gone so far into itself that it has nowhere else to go but back out, sending out new lines of flight. It turns out that in order to survive and thrive, synthetic lives such as minimal genome organisms need to be able cooperate with one another and build themselves back up again, molecule by molecule, in an environment-dependent and context-ridden "natural" world. Computational biologist Wenying Shou and colleagues argue that in the context of synthetic biology, "cooperative interactions are key to diverse biological phenomena" and that "such diversity makes the ability to create and control cooperation desirable for potential applications in areas as varied as agriculture, pollutant treatment, and medicine."[72]

Recognizing the importance of cooperation, Shou and colleagues show that "persistent cooperation can be engineered."[73] They state: "Specifically, we report the construction of a synthetic obligatory cooperative system, termed CoSMO (<u>co</u>operation that is <u>s</u>ynthetic and <u>m</u>utually <u>o</u>bligatory), which consists of a pair of nonmating yeast strains, each supplying an essential metabolite to the other strain. . . . Extending synthetic biology from the design of genetic circuits to the engineering of ecological interactions, CoSMO provides a quantitative system for linking processes at the cellular level to the collective behavior at the system level, as well as a genetically tractable system for studying the evolution of cooperation."[74] Linear thinking and reductionism have run their course, bringing us full circle. It turns out that synthetic organisms themselves are asking scientists to consider Butler's question, "What of life exceeds the model?"[75] Answering this call presents an opportunity for feminist scientists and feminist STS scholars. It invites us to consider what our responses will be, and what our encounters with these organisms will look like, when we realize that synthetic lives become expressive lives, capable of developing the quality of changefulness and desires for kinship.

Alloplastic Stratum: Deterritorializations through Postcolonial and Decolonial STS

In *Life as Surplus: Biotechnology and Capitalism in the Neoliberal Era*, Melinda Cooper examines how the biotech revolution of the 1970s and early 1980s shifted economic production to the genetic, microbial, and cellular level.[76] She argues that the transformation of biological life, including bacterial life, into surplus value is at the core of the new postindustrial economy. I am interested in extending Cooper's astute analysis of bacterial life and labor and other social, political, and economic factors to our own machinic assemblage. Postcolonial and decolonial STS can help to reframe synthetic biology along the social or alloplastic stratum. In particular, I am interested in using postcolonial and decolonial STS analyses to ask how, for example, are humans and nonhumans being organized to "manage the problems" posed by synthetic lives?[77] How, and from where, are the vast amounts of biomass that are required to support synthetic lives being obtained? The last phase of the synthetic life cycle progresses to a social phase that requires a great deal of support from both human and nonhuman systems and resources. New forms of labor and production are emerging as a result of these synthetic lives. I am interested in tracing those stories that shed new light onto our machinic assemblage, which up until this point has been comprised of various components, including humans, machines, digital DNA, minimal genomes, bacteria, yeast, and surrogate cells. As I explore this stratum, I analyze our machinic assemblage for its monstrous couplings with an STD, sugarcane plantations, and the Sargasso Sea.

The postcolonial and decolonial STS projects of thinking about "knowledge that is otherwise" and "reframing" biotechnological events resonate with new lines of flight that can form within Deleuze and Guattari's social or alloplastic stratum. In particular, the goal to actively "decolonize relations and practices" works hand in hand with Deleuze and Guattari's idea of deterritorialization.[78] Using the alloplastic to describe social institutions and behavior that are human but not limited to the human, Deleuze and Guattari write:

> There is a third major grouping of strata, defined less by a human essence than, once again, by a new distribution of content and expression. Form of content becomes "alloplastic" rather than "homoplastic"; in other words, it brings about modifications in the external world. Form of expression becomes linguistic rather than genetic; in other words, it operates with symbols that are comprehensible, transmittable, and modifiable from outside. What some call the properties of human beings—technology and language, tool and symbol, free hand and supple larynx, "gesture and speech"—are in fact properties of this new distribution.[79]

The alloplastic therefore is seen as a social or cultural stratum that creates new forms of content and expression. Aligning our analysis of a technology along the alloplastic stratum can be useful to understand how a machinic assemblage is working to modify the external world. In the case of content, we have those monstrous couplings that bring together several different kinds of physical bodies. In the case of forms of expression, we have new forms of technology, tools, and language used by humans (but not limited to humans) that also work to modify the external world that are a result of similar assemblages.

For example, we can begin to map those sides of the machinic assemblage that face new economies of biocapital that have become possible through the labor and protein-production capacities not only belonging to minimal genomes organisms but also to those humans whose labor supports synthetic life. We can begin to take account of how this labor can be contextualized along colonial histories of plantation-based economies or recent forms of biopiracy. We can begin to approach new forms of expression created by this machinic assemblage through critiques of scientific imperialism and liberal humanist notions of individualism as they relate to synthetic biology. Since it is also in the alloplastic stratum that expression becomes most independent from content, allowing for the greatest amount of deterritorialization, we are further able to contextualize, resituate, know otherwise, and reframe these events through other symbolic means, such as through the expressivity found in our stories and literature.[80]

Vignette 1: STDs

To begin, we can examine how the minimal genome organism brings with it a new genesis story. In January 2008 a team of seventeen scientists at the J. Craig Venter Institute (JCVI) in Rockland, Maryland, announced that they had successfully created the first synthetic bacterial genome.[81] Using the top-down approach and a variety of genetic engineering techniques, including *in vitro* recombination, cloning, PCR, *in vivo* recombination in yeast, and "shotgun" sequencing, Venter and his colleagues synthesized, assembled, and cloned the complete bacterial genome referred to as *Mycoplasma genitalium* JCVI 1.0. In 2016, Venter and his colleagues produced the even more streamlined version of the minimal bacterial genome referred to as JVCI-syn3.0, which contains only 531 kilobase pairs coding for 473 genes.[82] Interestingly, members of the Action Group on Erosion, Technology, and Concentration (ETC), an organization that analyses the socioeconomic ramifications of new technologies and is dedicated to the sustainable advancement of ecological diversity, referred to JVCI-syn1.0 as the "original syn." They have since dubbed JVCI-syn3.0 as Synthia 3.0.[83] Given Venter's previous ventures, we should have known that this day was coming.

In 1984, Venter held a position at the National Institutes of Health, where he began to work on a new technique for rapid gene discovery. He takes credit for developing a DNA sequencing technique referred to as expressed sequence tags (ESTs). In his biography on the JVCI website, it is suggested that in 1995 the ESTs technique led him to decode the genome of the first free-living organism using his new whole genome shotgun technique.[84] This was not the end for Venter and his biotechnological ambitions.[85] In fact, Venter actually traces his move toward synthetic biology to 1995, when he sequenced *Haemophilus influenza*. This genome was found to have about 1,800 genes. The same year, Venter collaborated with other scientists to work on the bacterium *Mycoplasma genitalia*. This bacteria was chosen because it has the smallest "naturally occurring" genome of any self-replicating organism, with only about 482 protein coding genes and approximately 580 kilobases.[86] However, some of us may find it extremely interesting to know that *Mycoplasma genitalia*, the "original" organism from which Venter's transformed minimal bacterial genome organisms are based, is a bacteria that causes a sexually transmitted disease (STD) in humans, known to lead to pelvic inflammatory disease (PID) in women. PID is a "major public health problem associated with substantial medical complications (e.g., infertility, ectopic pregnancy, and chronic pelvic pain) and healthcare costs."[87] In men, *Mycoplasma genitalia* is the "third most frequent pathogen causing non-chlamydial, non-gonococcal urethritis."[88]

Why would Venter choose a bacterial organism, known to cause a debilitating human disease that likely affects already vulnerable populations disproportionately, to serve as the biological backbone for the first synthetic life? From postcolonial and decolonial perspectives, we can see neoliberal and capitalist strategies at work. In this business model the concern over whether a minimal genome organism derived from a STD-causing bacteria poses a health concern to already economically and politically vulnerable groups is overshadowed by the speculative futures promised by synthetic biology. Referring to the growth of the pharmaceutical industry and the AIDS epidemic in Africa, Cooper explains that "one could go further along these lines and argue that the simultaneity of the North American–led biotech revolution and the troubling return of infectious disease of all kinds, in both the developing world and advanced capitalist centers, is symptomatic of the intrinsic contradictions of capitalism. The peculiarity of capitalism on this argument would lie in its tendency to create both an excess of promise and an excess of waste, or in Marx's words, a promissory surplus of life and an actual devastation of life in the present."[89] As a modern technoscience of the global North, the possible futures that have been promised by synthetic biology include biotechnologies of personalized biomedicine, bioremediation, and bioenergy applications. These technologies are primarily geared toward already well-resourced groups, and apparently their potential benefits outweigh the risk associated with the possible spread of a minimal genome organism that is STD-adjacent.

Vignette 2: Sugarcane Plantations

In 2004 the Bill and Melinda Gates Foundation donated $42.6 million to fund Jay Keasling's research on developing a synthetic antimalarial drug.

A proponent of the pathways approach, Keasling and his partners at the biotech startup Amyris (a not-for-profit at the time) went to work using *E. coli* and brewer's yeast to design a microbial cell whose metabolic pathways could be manipulated to incorporate the production of artemisinic acid, the precursor of the compound artemisinin.[90] Originally extracted from the plant *Artemesia annua* found mostly in China and southeast Asia, artemisinin has been the favored antimalarial drug for several years now due to the fact that plasmodium parasites have become resistant to quinine- and chloroquine-based treatments.[91] The reported problem back in 2005, when Keasling was conducting this research, was that plant-derived artemisinin was in "short supply and unaffordable to most malaria sufferers."[92] The tools of synthetic biology were supposed to fix this problem.

In 2006, Keasling and his team reported that they were successful in engineering the yeast *Saccharomyces cerevisiae* to produce high titers of artemisinic acid through the process of fermentation.[93] However, it was not until 2013 that they were able to report the production of "commercially relevant concentrations" of artemisinic acid.[94] The key limiting factor had been the ability to sustain the growth and fermentation of the synthetically modified brewer's yeast at an industrial level.[95] Since then, Amyris has become a private for-profit company, and using the technologies and expertise gained by having to produce semisynthetic artemisinin at industrial levels, they have expanded the applications of their synthetic microbial engineering model to include mass-scale production of cosmetics and biofuels. They have designed a synthetic yeast cell to produce high levels of the molecule farnesene, which "has many potential applications as a renewable feedstock for diesel fuel, polymers, and cosmetics." [96] Fermentation in yeast species such as *Saccharomyces cerevisiae* is a metabolic process that converts sugars and starches into acids, gases, and alcohol. However, in order to carry out industrial levels of yeast fermentation, one also needs industrial amounts of sugar and starch-based biomass for the desired metabolic processes to occur. Therefore, while entering the market of cosmetics and biofuel production, Amyris also purchased sugarcane fields in Brazil to carry out their mass-scale operations.

As a global leader in biofuel production, Brazil has been producing ethanol-based biofuel from sugar and sugarcane-derived biomass for several decades. Amyris decided to develop its own farnesene manufacturing facilities by using the country's already well-established sugarcane production infrastructure. As such, the company acquired portions of existing sugarcane fields as well as new feedstock facilities that are adjacent to existing sugarcane mills, such as in the municipality of Brotas, in São Paulo, Brazil. Adrian MacKenzie explains:

> The Brazilian sugar-cane industry is the largest producer of sugar in the world. Rather than producing ethanol through the long-established industrial techniques of fermentation, some of the Brazilian sugar-cane will become something different in Amyris' bolt-on bioreactors at Usina São Martinho in Brazil. The years of metabolic engineering that Keasling's team put into the isoprenoid pathway in yeast pays dividends now in the form of a usefully transformable chemical, farnesene. The millions of tons of sugar cane moving through Usina São Martinho no longer simply ferment as ethanol, the biofuel that Brazil has produced in quantity since the 1970s. Via Amyris' re-engineered yeast strains, the chemical substrates present in sugar will be re-routed as feedstock for a much more complicated and efficient metabolic pathway, the melavonate or "HMG-CoA reductase pathway."[97]

Drawing on the philosophical work of Gilbert Simondon, MacKenzie conducts a rich analysis of biofuel, treating it as a technical object whose "genesis involves processes of concretisation that negotiate between heterogeneous geographical, biological, technical, scientific, and commercial realities."[98] I would add to this list of realities the colonial histories of sugarcane plantation–based capitalist economies, the indigenous peoples who were displaced or killed by European settlers, and the labor and bodies that were organized by this economy.

Brazil, like many other countries in the Caribbean and Latin America, experienced its first wave of European colonial expansion soon after Christopher Columbus returned from his initial voyage to the "new world." In fact, the earliest record of large-scale sugar production goes back to 1550, when the Portuguese built mills along the Atlantic coast of Brazil.[99] Caribbean scholar Fernando Ortiz discussed the politics of tobacco and sugar production in Cuba in his influential work *Cuban Counterpoint*, and his analysis of the European establishment of sugar plantations as a strategy for economic claims to the colonies can be extended to Brazil. Ortiz explains:

> It is one thing to have cane and another to produce sugar on a commercial scale. Between the raising of the cane, which experience had shown to be merely a question of man power, and the commercial production of sugar, which in Europe had a steady and growing market, stood the problem of industrial production, which demanded machinery and technicians that did not exist here, and of necessity had to be imported from Europe. In a word, capital was needed to buy slaves, to bring in experts and skilled workers and all the machinery for milling, boiling, evaporating, and refining. Even aside from the land required, the production of sugar was perforce a capitalist enterprise.[100]

Postcolonial and decolonial analyses encourage us to probe what the effects of sugarcane production were not only on the local indigenous populations in Brazil but also on the slaves who were brought from Africa over a period of roughly three hundred years to sustain the industrial production of sugar. What modifications to the external world were caused by the machinic assemblage that was, at the time, a monstrous coupling of sugarcane plantations, mills, slaves, and sugar? What were its effects on the lives of individuals who had been displaced by slavery? Ortiz describes the effects of displacement and the backbreaking labor in those sugar plantations:

> At the same time there was going on the transculturation of a steady human stream of African Negroes coming from all the coastal regions of Africa along the Atlantic, from Senegal, Guinea, the Congo, and Angola and as far away as Mozambique on the opposite shore of that continent. All of them snatched from their original social groups, their own cultures destroyed and crushed under the weight of the cultures in existence here, like sugar cane ground in the rollers of the mill . . . [they] brought with their bodies their souls, but not their institutions nor their implements. . . . They arrived deracinated, wounded, shattered, like the cane of the fields, and like it they were ground and crushed to extract the juice of their labor.[101]

Given the wealth that was generated by slave labor in the sugarcane plantations, it is no surprise that Brazil was the last country to abolish slavery in the Americas.

Although I am no way suggesting that Amyris's new sugarcane mills that produce semisynthetic cosmetics and biofuels employ slave labor in their fields, maintaining the sugarcane crops and operating the fermentation plants must involve the extraction of local human labor. Even if much of the processes are now mechanized, a machinic assemblage that brings together a monstrous coupling of sugarcane fields, mills, laborers, sugar, and synthetic yeast organisms is still an assemblage that modifies the external world—namely by organizing the bodies and cultures of specific humans. It is important that the history of labor practices and worker conditions associated with sugar and sugarcane-derived biomass in Brazil and other countries not be forgotten.[102] These histories can be used to better understand the effects incurred by the practices of transnational companies such as Amyris and the naturalization of similar capitalist practices. For instance, in her efforts to create anticapitalist transnational feminist practices, Chandra Talpade Mohanty asks us to bring forward the question of native or indigenous struggles in our analyses. "Economically and politically," she writes, "the declining power of self-governance among certain poorer nations is matched by the rising significance of transnational institutions such as the World Trade Organization and governing bodies such as the European Union, not to mention for-profit corporations. . . . [T]he hegemony of neoliberalism, alongside the naturalization of capitalist values, influences the ability to make choices on one's own behalf in the daily lives of economically marginalized as well as economically privileged communities around the globe."[103]

As we think about the sugar and sugarcane biomass needed for the production of malaria drugs, cosmetics, and biofuels by synthetically developed microorganisms, we should keep in mind that our decisions regarding the development and commercialization of new technologies and products for consumption will have local and global impacts. We must also keep in mind that many of these decisions are being made without input from the people whose lives will likely be disproportionately impacted. As ETC spokesperson Jim Thomas has pointed out, we live in an unjust world and if we want to develop technologies that are not going to add to this injustice, synthetic biologists need to realize that marginalized communities must have a say in what comes to constitute their reality.[104]

Vignette 3: The Sargasso Sea

Interdisciplinary scholarship is a fantastic place for discovery, but sometimes ideas can easily get lost—a bit of a scholarly Bermuda Triangle one might even say. With this admission, I end this chapter by moving quickly into muddier waters, to the Sargasso Sea in particular, with the hope that our machinic assemblage doesn't get marooned. Despite its reportedly weak currents and calm winds, the Sargasso Sea located in the North Atlantic Ocean is a busy place, playing host to the imaginations of colonial explorers, novelists, postcolonial theorists, marine microorganisms, pirates, biopirates, and synthetic biologists. Like the floating beds of sargassum seaweed, after which the sea is named, entanglements come easily here. Here, our machinic assemblage is oriented to face the alloplastic stratum, where postcolonial and decolonial perspectives help us to reflect on synthetic life and the emergence of neoliberal forms of individualism and imperialism.

In 2004, J. Craig Venter and colleagues published the article "Environmental Genome Shotgun Sequencing of the Sargasso Sea" in the prestigious journal *Science*. Using his personal yacht, the *Sorcerer II*, Venter and his team had taken sail a few years earlier and applied the whole-genome shotgun sequencing technique to "microbial populations collected en masse . . . from seawater samples collected from the Sargasso Sea near Bermuda."[105] Funded largely by the US Department of Energy as well as the Discovery Channel, Venter and his team of scientists set sail again aboard the *Sorcerer II* in 2009, this time with the intention of traveling around the globe, collecting more marine microbial samples.[106] Why this interest in marine microbes? Similar to the scientists at Amyris, Venter and many others saw the promise of using synthetic organisms to produce biofuels. However, instead of using brewer's yeast and *E. coli*, in Venter's case the synthetic powerhouse he had in mind for the job of biofuel production was a marine microbe, particularly a microalgae. This microalgae-based future was full of so much promise that in 2009 the oil and gas giant ExxonMobil partnered with Venter's startup Synthetic Genomics Incorporated and contributed $600 million to jumpstart synthetic biofuel research.

Since the 1970s, scientists have been experimenting with different strains of microalgae to take advantage of their ability to produce lipids, which can be transformed into biofuels. The main stumbling block for these scientists has involved finding a microalgae that can photosynthesize "efficiently" enough to convert light energy and CO_2 into industrial levels of biomass and lipid production.[107] In 2017 it was announced that the collaboration between ExxonMobil and Synthetic Genomics had finally led to the creation of a synthetically engineered and phototropic strain of microalgae (*Nannochloropsis gaditana*) that would overcome these barriers. As the scientists of this joint venture explain in an article published in *Nature Biotechnology*, they developed a CRISPR-Cas9 reverse-genetics pipeline and used it to identify and modulate expression of a lipid regulator in *N. gaditana*, increasing its lipid production to commercially relevant levels. They state: "Using a microalga to produce lipids offers the potential advantages of being able to phototropically convert CO_2 to lipids without relying on agriculturally derived sugars, thus mitigating the demand for arable land and freshwater. Our findings represent a step toward understanding and controlling lipid production in algae. This ability to control algal lipid production might eventually enable the commercialization of microalgal-derived biofuels."[108] Cofounder, chairman, and co–chief scientific officer of Synthetic Genomics, Venter adds that "the SGI-ExxonMobil science teams have made significant advances over the last several years in efforts to optimize lipid production in algae. This important publication today is evidence of this work, and we remain convinced that synthetic biology holds crucial answers to unlocking the potential of algae as a renewable energy source. We look forward to continued work with ExxonMobil so that eventually we will indeed have a viable alternative energy source."[109]

While we learn about the promises of synthetic algae serving as a source of biofuel that meets our growing needs for alternative energy sources, we also witness the intimate partnering or placing together of synthetic biology research and a giant transnational oil and gas company.[110] Some of us may be reminded of the Exxon *Valdez* oil tanker spill in 1989 and the environmental disaster that occurred in the waters off the coast of Prince William Sound, Alaska. Some of us may also remember the devastating effects of this environmental disaster on local wildlife and the economic effects that were felt by local indigenous communities in the area. Postcolonial and decolonial STS analyses urge us to recall this history as we attempt to reframe current microbe exploration and synthetic biofuel research that is being supported by ExxonMobil. We may begin to see a pattern of imperialist practices that include the impulses to explore "uncharted" spaces and to claim direct access to or ownership over natural resources in a faraway land—biopiracy by other means.

In this light, the vision of Venter sailing in his luxury yacht in the waters of the Sargasso Sea, collecting samples of microalgae to sequence their DNA, begins to look less like a journey of basic scientific inquiry into the evolution of marine life and more like the voyage of a venture capitalist who is shoring up promissory futures by gathering DNA samples and data from microbial life forms. Although patent applications were submitted for his two minimal genome organisms, Venter claims that he is not interested in patenting the microbial genome sequences obtained from the Sargasso Sea. Indeed, the strategy that he and others have developed is not to place a patent on the "natural" genome sequence of an organism itself. This information is entered into public databases such as the National Institutes of Health's GenBank. Rather, a patent is taken out on the tools and technologies developed to design and engineer synthetic genome organisms based on these "natural" organisms.[111]

Venter has gone so far as to chastise the governments of poorer countries who share the waters of the Sargasso Sea such as Bermuda for voicing concern over Venter's collection of water and soil samples from within their national coastal borders. In an interview with *Discovery Magazine*, Venter stated:

> Most of the ocean is claimed by one or more countries. A lot of politics is building up around this thing. So now we're evil because we're putting data in the public domain. A group of people who are following everything we do is putting out a lot of false information. You go on to some Web sites, and they say we're trying to patent everything. [Interviewer: Are you?] No, and that's the ultimate irony. We're doing the stuff and giving it to the world, and now it's evil because all these poor little countries like Bermuda want to profit somehow from this data. They don't realize that they can profit from the knowledge. I think people just like to attack what we're doing because we're always on the leading edge.[112]

It is unclear what the distinction is that Venter is trying to make by chastising so-called "poor countries" for wanting to profit from the data when

Venter's own privately held company Synthetic Genomics is poised to profit enormously from the commercial applications made possible by the very same data.

In an interview with *Bio-IT World*, Venter was reported to have said: "It was a big surprise to me that there's very little international waters left. I thought I was out sailing free in the ocean and somebody's claimed it all."[113] Venter's desire to sail free into the Sargasso Sea, and to sequence the genome of all the organisms he could find, paints a familiar picture of neoliberal forms of imperialism and individualism. In 2007, Venter took pride in this individualism while appearing on the talk show *The Colbert Report* to promote his book *A Life Decoded: My Genome, My Life*. Having unveiled his own genome sequence and made it publicly available on the internet, Venter said, "I think we found that we're far more different than each other than we thought even a few years ago. We're 1 to 2% different instead of one letter out of a thousand base pairs. We don't all have the same genes—we have major differences. As an individualist, I find that very encouraging."[114] Venter sees himself as an individualist and is delighted to have discovered that humans are more different from each other, by an entire order of magnitude. Basically, he is happy to announce that he is even more different from his human others than he previously thought. This understanding of difference is motivated by individualism and is not the proliferative difference found in Deleuze's ontology of univocity. Postcolonial and decolonial STS analyses would have us consider what the consequences will be of this newer version of an old worldview, whereby the white male human subject gets to distance himself even further from his human others, let alone the nonhuman synthetic others that he creates.

Years ago, in her essay "Three Women's Texts and a Critique of Imperialism," Gayatri Spivak also took us on a voyage to the Sargasso Sea. Writing about the crucial role of imperialism in the "cultural representation of England to the English," she directed us to the novel *Wide Sargasso Sea* (1966) written by Jean Rhys, as well as Charlotte Bronte's *Jane Eyre* (1847) and Mary Shelley's *Frankenstein* (1818), to argue that the "project of imperialism has always already historically refracted what might have been the absolutely Other into a domesticated Other that consolidates the imperialist self." Spivak wrote this article mainly as a postcolonial critique of nineteenth-century English literature that promoted feminist individualism; however, she also pointed out that the continued success of the imperialist project is due to it being "displaced and dispersed into more modern forms."[115] When Spivak published this piece in 1985, I don't think she was thinking about synthetic biology, minimal genome organisms, or the production of biofuels by nonhuman, nonanimal, marine microbes in the Sargasso Sea as modern forms of the imperialist project. But who knows, perhaps she was, for the imperialist project that Spivak describes in the three novels above speaks surprisingly well to the neoliberal practices being co-constituted with synthetic biology today.

For instance, in *Wide Sargasso Sea*, Rhys tells us the story of a white Creole woman from Jamaica. The novel delivers the untold story of Bertha Mason, the "mad" West Indian woman in Bronte's *Jane Eyre* who is locked up in her husband's house in England but in the end manages to escape, burn down the house, and take her own life. Born in Dominica in 1890, and being of white Creole descent herself, Rhys saw the injustice being played out in Bronte's literary treatment of Caribbean women through the character depiction of Bertha. *Wide Sargasso Sea* is written as a prequel to *Jane Eyre* and tells the heart-wrenching story of Antoinette (renamed Bertha) Mason's life, with a backdrop in a time and place when the emancipation of slaves was under way in the British colonies of both Jamaica and Dominica. Rhys's novel, in its own way, addresses the cost of British imperialism on human lives by attempting to give a voice to the other within Bronte's text. In her critique of *Wide Sargasso Sea*, however, Spivak suggests that the story of Antoinette, presented as the unwritten story of a white Creole woman, in fact "reinscribes" the weighty absence of the other in Bronte's *Jane Eyre.* "As the female individualist, not-quite/not-male, articulates herself in shifting relationship to what is at stake," Spivak writes, "the 'native female' as such (within discourse, as a signifier) is excluded from any share in this emerging norm."[116]

Much in the same vein that Antoinette or Bertha's story is absent in *Jane Eyre*, Spivak notes that "Christophine's unfinished story is the tangent" in Rhys's *Wide Sargasso Sea*.[117] In the novel, Christophine is a black woman from Martinique who practices obeah, was given to Antoinette's mother as a wedding gift, and serves as Antoinette's nurse. Although

Christophine's character is given a crucial role in the novel, much like Bertha in Bronte's text, Christophine disappears in *Wide Sargasso Sea* after confronting Antoinette's husband, who embodies the mission of British imperialism. There is a distinction to be drawn, however, in the disappearance of Christophine and the limited depiction of Bertha. As Spivak suggests:

> She [Christophine] cannot be contained by a novel which rewrites a canonical English text within the European novelistic tradition in the interest of the white Creole rather than the native. . . . Attempts to construct the "Third World Woman" as a signifier remind us that the hegemonic definition of literature is itself caught within the history of imperialism. A full literary reinscription cannot easily flourish in the imperialist fracture or discontinuity, covered over by an alien legal system masquerading as Law as such, an alien ideology established as only Truth, and a set of human sciences busy establishing the "native" as self-consolidating Other.[118]

Spivak points out that although such novels as *Wide Sargasso Sea* and *Jane Eyre* are often celebrated as proto-feminist works, they also promote forms of feminist individualism that feed into the imperialist project. Through these literatures we can see how such individualism was only able to occur through the subjugation of others. Spivak is critical of the glorification of feminist individualism, for she argues that "what is at stake . . . in the age of imperialism, is precisely the making of human beings, the constitution and 'interpellation' of the subject not only as individual but as 'individualist.'"[119] In an era of biocapital, the self-interest of individualism and the legacy of the imperialist project collide in fascinating ways. Of course, it goes without saying that the self-interest of the individualist gels very nicely with the subject formation necessary to propel oneself forward in a global economy based on such neoliberal values. It is this very skill set that is required for one to venture and set sail, as Venter and others have done, into a life of biopiracy.

This chapter has covered a great deal of ground (and water). I have interrogated two main challenges we will face as we turn our attention toward the *in vitro*, the molecular, the synthetic, and the nonhuman. First, we must consider the question of what constitutes life and the living, and second, we must figure out how to orient these questions so as not to leave the human behind. Interdisciplinary incubations produced through the encounters between molecular biology, feminism, postcolonial theory, and decolonial studies will be key in addressing these challenges. As we learn how to see the world and furthermore, how to encounter that world, these incubations will sustain us in our pauses even as we race toward new biotechnological futures.

Conclusion

Science in Our Backyards

| grass · root |
the root of a plant of grass

—*OXFORD ENGLISH DICTIONARY*

| grass · roots |
the most basic level of activity or organization

—*OXFORD LIVING DICTIONARIES*

The term "grassroots" is a plural noun that has found its way into all sorts of political spaces. It seems that these days everyone is organizing in some way or another at the grassroots level. Grassroots movements range from those that work tirelessly to save the planet, those that travel door-to-door campaigning for a political candidate, those that strive to end gendered violence, to those that work fearlessly against racial and social injustices, such as Black Lives Matter. It is interesting to think about the properties or qualities that actually make an organization or movement into one that is considered to be "grassroots." If, as it is defined by the *Oxford Living Dictionaries,* for something to qualify as being "grassroots" it must meet the most *basic level* of activity or organization, then the question becomes, Who or what counts as being "basic"? Who or what decides how we discern what constitutes a "level"? What boundaries and borders are we dealing with here?

For instance, starting as a call to action against state-sanctioned violence and anti–black racism, Black Lives Matter proudly stands as a leaderless social justice movement that is instead "member-led." Does being "member-led" count as being a grassroots organization or an activity that takes place at the most "basic level"? The movement does affirm several key ideas and aspirations that serve as the organization's guiding principles. As one of the most powerful movements of our time, Black Lives Matter encourages its members to work toward and affirm values of diversity, restorative justice, globalism, collective value, empathy, and loving engagement. The movement is intergenerational, queer and transgender affirming, and unapologetically black, affirming black villages, black women, and black families.[1] As inspiring as these principles are, how is it that a member-led organization—one that is comprised of a global network—can aspire to build upon so many rich, but also disparate, principles? How does a grassroots movement such as this take action?

Many grassroots movements, including Black Lives Matter, are built on the idea that everyday people can come together to organize and effect change. This indeed is the beauty and hallmark of doing things the grassroots way. Yet many such "member-led" movements commonly face criticisms for being "horizontal" rather than "vertical" in their organizational structure—that is, without a clear hierarchy in their leadership. In their contribution to the anthology *Urban Policy in the Time of Obama*, Lorraine Minnite and Frances Fox Piven have commented on the capacity of horizontal movements to exercise power. They explain that a "vertical" organization is mainly defined by unequal social relations of status and class, and that although they suggest that there is a chance for those at the bottom of this vertical organization to access power, this access mostly comes from "locating vulnerabilities" within a hierarchical structure.[2] Minnite and Piven go on to say that those at the bottom of a vertical organization can still effect change by "scale jumping" or, in other words, by overcoming institutional hierarchies that "lock in place the deep inequalities of urban life."

Alternatively, they explain that horizontal urban social movements work as lateral networks of support.[3] For these lateral networks to function, they must overcome several boundaries and borders by building coalitions through common interests. Interestingly, however, Minnite and

Piven suggest that in both the cases of vertical and horizontal organizational structures, change is produced through "disruptive action." They explain: "In this sense, movement activities and actions do not so much 'jump scale' . . . ; rather, the effects of their actions pulse like electricity through the nodes of networked relations, both horizontal and vertical, the energy reconfiguring those relations to shrink the spaces in between, and bend authority in a favorable direction." The definition of "disruption" that Minnite and Piven are working with here relates to urban social movements and refers specifically to "actions that withdraw contributions to social cooperation within institutional arrangements."[4]

Yet, as members of the Black Lives Matter movement would present the results of their collective efforts, it appears that the definition of what gets to count as a "disruptive action" in an urban movement is far more varied than simply withdrawing one's contribution to institutional arrangements. In fact, if one goes onto the Black Lives Matter website, they will find a link called "Take Action." If one follows that link, further options to either "Find an Action" or "Take an Action" are presented. According to the movement, the disruptive actions that have been taken by their members have led to a number of accomplishments, having "ousted anti-Black politicians, won critical legislation to benefit Black lives, and changed the terms of the debate on Blackness around the world."[5] It would appear that this grassroots, member-led movement, which is supposedly operating at the "most basic level," can accomplish extremely complex tasks and effect change at many different ways.

This book is not about urban social movements. Yet it is predicated on the idea that by the mere act of pronouncing "molecular feminisms," the reader will have engaged in several disruptive actions. The first disruptive action comes from the fact that the two fields, molecular biology and feminism, have been kept apart. For many reasons, including institutional structures and disciplinary gate keeping, the two fields rarely have a chance to "shrink the spaces in between" them. A second disruptive action of molecular feminisms follows from the fact that although this book brings together the topics of feminism and science, it does not do so by turning to predominant or majoritarian discourses of gender equality or banal pipeline metaphors that usually accompany most women and science projects. It reaches instead toward those minor literatures or less explored lines of inquiry that are invested in questions of ontology, ethics, epistemology, and everything else in between that goes into the political act of knowledge-making. Lastly, thinking with molecular feminisms serves as a disruptive action because at the heart of this project is precisely the idea that molecular biology itself can provide us with creative ways to enact critical disruptions and thereby reconfigure dominant relations.

Discussing urban social movements, Minnite and Piven describe the effect of movement activities and actions as a "pulse of electricity" that "moves through both nodes and networks" in order to reconfigure relations. It is here, in their description of *how a disruptive action works*, I realize that in writing a book on molecular biology, feminist theory, Deleuzian philosophies, postcolonial theory, and decolonial studies, I might have been trying to describe strategies of horizontal social movements all along. The only difference is that what counts as "social" in my case has not been limited to the human experience. It has included not only humans but all nonhumans, including organic and inorganic others. My understanding of the social has even included "raw" matter. Molecular feminisms, biophilosophies of becoming, and microphysiologies of desire all attempt to reframe these social relations by turning to ontological and ethical maneuvers that create movement and disruptive actions along a horizontal plane.

These social relations include those between the humanities and sciences, culture and biology, feminist theorists and feminist scientists, and between the knower and what is to become the known. To reframe these relations, I have taken us back to the actual physical and material pulses of electricity that Jagadish Chandra Bose reported to have measured when he was conducting his experiments on the capacity for response in plants. Developed by an anticolonial figure, who conducted scientific research as a colonial subject under British rule, I have used Bose's work to theorize a different understanding of "response" that may be useful to both feminism and biology. Similarly, I have brought forward the work of such scientists as Barbara McClintock and Lynn Margulis, who may not have referred to themselves as feminists but whose intimate scientific inquiries of chromosomes and bacteria can serve to address some of the most pressing questions in feminist theory and feminist STS today.

Each of these scientists—Bose, McClintock, and Margulis—were marginalized within their scientific communities for different reasons in their day. Yet their theories and research findings have lived on and now contribute to our efforts to bend scientific authority and to produce knowledge that is otherwise. My reason for sharing their scientific work has been the hope that by learning about their disruptive actions, we as feminist scientists, scientist feminists, and all other invested parties may also be called to action. Like the strategies of horizontal urban social movements, much of our efforts actually need to start locally, in our own backyards as it were. Practice-oriented feminist STS approaches can help in this regard by providing us with the everyday knowledge and tools to conduct our experiments. Doing science in our backyards could include setting up local science shops where experts from a diverse range of knowledge bases come together, through a shared common interest, to solve local problems. It could involve creating feminist, postcolonial, and decolonial technoscience salons where academics learn to bring their research into interdisciplinary conversations. It could involve creating shared community maker spaces that prioritize the involvement of typically marginalized groups. It could even involve setting up interdisciplinary mentoring structures that support the radical act of having feminist scientists practice both their science and their feminism at the lab bench.

Lastly, I want to address the first definition of "grass root" placed at the start of this conclusion. According to the *Oxford English Dictionary*, a "grass root" is a noun that can be defined simply as the "root of a plant of grass." Yet, if I have succeeded in any way at all in writing this book, it should be evident by now that the coming together of stolons, rhizomes, roots, shoots, and molecular politics in the event that is *becoming a blade of grass* is anything but simple or inconsequential. I learned this lesson myself not from arboreal thought but from a tree that fell in my backyard.

Glossary

agrostologist: A scientist who studies the branch of botany concerned with grasses.

alloplastic: A term used by Gilles Deleuze and Félix Guattari to describe the social and cultural stratum.

anticolonial: A position held by a person or a country that opposes colonialism.

applied ethics of matter: A way of putting one's ontological and ethical positions into action.

assemblage: An "assemblage" holds a heterogeneous collection of components or elements together through complex sets of relations. It is also an ontological framework developed by Gilles Deleuze and Félix Guattari to understand the multiple couplings and various combinations of bodies, expressions, institutions, and signs.

becoming: This philosophical concept emphasizes the capacities of flux and change, where the properties and components of nature are understood in terms of process and events.

bench science: Scientific research that is conducted in a laboratory at a bench.

biophilosophies of becoming: An ontological approach to thinking about the properties and components of nature as they relate to our research in biology and our approach to biological matter.

bodies without organs (BwO): A phrase used by Gilles Deleuze and Félix Guattari that is not a stance against organs themselves. It is an attempt to think in a way that does not organize and produce boundaries around objects, but rather opens up our thought to the possibilities of broader connections and intensities (dis-organ-ization).

continental philosophy: Philosophical traditions that emerged in the nineteenth and twentieth centuries out of Europe.

cosmopolitics: A term used by Isabelle Stengers to describe an approach that does not avoid risks while working toward an experimental togetherness between two different communities of knowers.

decolonial/decolonizing: Terms initially used by an emerging Latin American scholarly movement that applies critical theory to and from within ethnic studies to think about modernity and coloniality.

deterritorialization: The act of thinking otherwise and questioning those modes of analysis that claim to hold a position of organized or authoritative knowledge.

epigenetics: Literally meaning "that which is above the gene," epigenetics is the study of how heritable traits can be modified by nongenetic and/or environmental factors.

epistemology: The theory of how we know what we know.

essentialism: The belief that there is a real and invariable nature at the root of any given thing.

estrogens: A group of steroid hormones derived from cholesterol that plays a role in growth, development, and reproductive functions.

ethico-onto-epistemological: A term used by Karen Barad to describe the simultaneous events of learning how to see the world, learning how we come to know the world, and thinking about how we learn to approach that world.

feeling around for the organism: The book puts forward this approach as a way to think about our encounters with the world. It belongs to a group of microphysiologies of desire proposed throughout the book that serves as an applied ethics of matter. It is a way to think about the relationship between the knower and the known.

haecceity: A Latin-derived term that Gilles Deleuze and Félix Guattari use to describe an event that exists on a plane of immanence.

humanism: A system of thought whereby humans as well as their actions, values, and interests are placed at the center of inquiry.

hylomorphism: A tradition of philosophical thought that views matter as being passive or inert until it assumes a pre-given form.

hylozoism: A tradition of philosophical thought that recognizes the expression of certain capacities in all forms of matter.

immanence: Central to Gilles Deleuze's ontology, this concept puts forth the idea that all becomings exist beside or with other becomings upon a horizontal plane of immanence. This concept is contrasted with the concept of transcendence, which Deleuze associates with a Platonic and hierarchical distinction between matter and form and with fixed or essentialized traits.

immunostaining: An antibody-based method used to detect a specific protein in a sample.

intra-action: This term, used by Karen Barad, describes the mutual constitution of agents that are always entangled with each other.

***in vitro*:** A process that takes place outside of a living organism or body. Throughout this book *in vitro* refers to scientific research that takes place within cell cultures rather than in whole animal models.

kinship: A relation made possible by developing connections and ties with others. Rather than seeing genetic or blood relations as the only way to form kin, feminist scholars have expanded the scope of who and what we are able to connect with and respond to and thus consider to be kin.

machinic assemblage: Similar to an assemblage, the machine assemblage is more about an approach to thinking rather than actual machines. It considers the monstrous couplings of heterogeneous components that go into the making of all actants.

major/majoritarian: Belonging to a well-established or dominant mode of thought.

materialism: Theories that deal with matter and/or material goods and place significance on their roles in constituting all phenomena.

metaphysics: This branch of philosophy deals with such ideas as cosmology, being (ontology), and knowledge (epistemology).

microphysiologies of desire: This term describes the approaches we use to move forward in our encounters with others. Microphysiologies of desire provide frameworks for putting our ethical positions into action. They are related to biophilosophies of becoming in the sense that ontology and ethics are also always related, if not the same thing. All microphysiologies of desire work by helping us to (1) cultivate an openness to nonhuman becomings and the capacity for changefulness; (2) make connections through kinship and sensing hylozoism; and (3) create movement by way of univocity and immanence.

micropolitics: Refers to those politics that do not follow our dominant habits or usual modes of engagement.

minimal genome organism: An organism whose genome has been reduced to a bare minimum number of genes required for growth and division of that organism.

minor/minoritarian: Belonging to a less explored or marginalized mode of thought.

molar: In chemistry this pertains to a solution containing one mole of a solute per liter of solution. A term also used by Gilles Deleuze and Félix Guattari to describe an ontology of being that is tied to fixed identities.

molecular: Relating to molecules, this term used by Gilles Deleuze and Félix Guattari describes an ontology of becoming that examines actants in terms of processes. Also used to describe minoritarian projects or politics.

monism: A concept that attributes a oneness to nature.

naturecultures: Coined by Donna Haraway, this term describes the commingling and co-constitutive properties of all things previously considered to be separate through such categories as nature and culture.

neuroendocrinology: A branch of the life sciences that studies the physiological interactions between the nervous system and hormones of the endocrine system.

onto-ethical: A term used to describe the idea that our theories on how to come to see the world are in and of themselves also always about questions of ethics, or questions about how to approach and treat that world.

ontology: A theory of being, or in the case of Gilles Deleuze and other philosophers, a theory of becoming. In other words, a theory of how we come to be and how we see the world.

Petri dish: A transparent, circular, and flat dish that is used to support cell cultures and/or study microorganisms in the lab.

plasmid: A small and circular double-stranded piece of DNA that is distinct from an organism's chromosomal DNA. Plasmids are generally found in bacteria.

postcolonial: Refers to the period following colonialism but also the study of the political, institutional, and cultural effects of colonialism and imperialism.

posthumanism: Although this term can be understood as meaning after or beyond the human, it is also a field of scholarly inquiry that is critical of humanism and attempts to decenter the human and human values and interest from our inquiries.

process ontology: A theory of being that is based on the idea that all beings are becomings and that there is a dynamic nature to all entities.

receptor: A protein structure that binds with substances such as hormones, drugs, or neurotransmitters.

rhizome: A continuously growing horizontal stem that grows underground and that can produce shoots and roots leading to the generation of new plants.

stolon: Also referred to as a runner in agricultural terms, the stolon is a stem system that grows above ground by developing new shoots and extending itself horizontally. The stems of a stoloniferous plant grow by sensing light and sending out aerial shoots. Other than cespitose grasses, all other grasses grow as rhizomes or stolons. Whereas rhizomes grow beneath the surface of the soil, stolons grow and move above ground.

strata/stratification: Terms used by Gilles Deleuze and Félix Guattari to describe a kind of grouping or a process of taking on particular forms and expressions.

subcloning: A recombinant DNA technology used to amplify a short segment of DNA by inserting it into a vector or plasmid that can be further replicated.

symbiosis: An evolutionary theory developed by Lynn Margulis that describes how the relationships between different organisms are the driving force of evolution.

synthetic biology: A field of research that brings together advances in biology, chemistry, genetics, computer science, and engineering to provide toolkits for the design of biological systems.

taxonomy: The science and practices of classification.

teleology: A direct or linear arc of events or reason used to explain a purpose or end goal.

transcendence: Although this term has several meanings, it is used here in the Deleuzian sense of being superior to or surpassing a level of existence and/or expression. Rather than immanence, which is understood as remaining with or beside something, "transcendence" is understood as being that is beyond or outside.

transcription: The process whereby DNA serves as a template for the production of messenger RNA.

transgenic: A term used to describe an organism that contains genetic material from another organism. Also describes the technologies used to introduce foreign genes.

translation: The process whereby messenger RNA is used to form a chain of nucleotides that in turn become a protein.

transpositions: The process of horizontal transfer of genetic materials between organisms. The process of "jumping genes" and the existence of transposable elements or transposons was discovered by Barbara McClintock.

univocity: A concept that everything shares a kind of oneness. This oneness does not mean that everything is identical or that there is no chance for differences, but rather that everything exists on a single or on a continuous ontological field.

Notes

Introduction: Stolonic Strategies

Epigraphs: Deleuze and Guattari 1987, 280; Kumar and Shivay 2008, 178.

1 Deleuze and Guattari 1987, 15.
2 Ibid., 15.
3 Kumar and Shivay 2008, 178.
4 Lorde 1984b.
5 Willey 2016.
6 This research took place in Denise D. Belsham's reproductive neuroendocrinology lab at the University of Toronto. I am grateful to Denise for her supervision, for sharing her expertise in molecular biology techniques, and for the cutting-edge research taking place in her lab.
7 See Roy 2007 and 2014.
8 Haraway 1991, 1997, 2004; Stengers 2000a, 2000b, 2005, 2010; Barad 2003, 2007.
9 Stengers 2010.
10 Hacking 1983, 150.
11 Myers 2015, xi.
12 Stengers 1997; Haraway 2008b.
13 Roy, Angelini, and Belsham 1999.
14 Yen 1991.
15 MacGillivray et al. 1998; Shevde et al. 2000; Baker et al. 2003.
16 Barakat et al. 2016.
17 Levine 1997; Roy, Angelini, and Belsham 1999; Belsham and Lovejoy 2005.
18 Shivers et al. 1983.
19 Kuiper et al. 1996.
20 Roy, Angelini, and Belsham 1999.
21 Leung et al. 2006.
22 Prossnitz, Arterburn, and Sklar 2007.
23 Terasawa and Kenealy 2012; Prossnitz, Arterburn, and Sklar 2007.

24 I first heard this term "scientist feminist" being used by my brilliant colleague Elizabeth Wilson, whose work has served as an important point of reflection in my own contributions to feminist STS.
25 D. Smith 1974; Rich 1976; Hartsock 1983; hooks 1984; Collins 1990; Haraway 1991; Code 1991; Harding 1991; Trinh 1991; Crenshaw 1993; Sandoval 2000, 2004; Alcoff 2006.
26 Grosz 1990, 337.
27 Ibid.
28 Ibid., 339.
29 Ibid.
30 For more information, see the Our Bodies Ourselves website at www.ourbodiesourselves.org/history/, the Feminist Women's Health Center website at www.fwhc.org/selfhelp.htm, and the Black Women's Health Imperative website at www.blackwomenshealth.org/.
31 The work of many different feminist scholars housed in several different disciplines came together to form these early interventions. For works by feminist philosophers during this era, see Harding 1986, 1987; Haraway 1988, 1990; Tuana 1989a, 1989b; Longino 1989, 1990; Hankinson Nelson 1990; Code 1991; Keller and Longino 1996; Rouse 1996, 2002; Wylie 2002, 2004; Potter 2001; Stengers 1997, 2000a, 2000b. For works by feminist historians of science, see Rossiter 1982; Keller 1983, 1985; Merchant 1990; Schiebinger 1991, 1994. For works by feminist sociologists and anthropologists of science, see Martin 1987; Star 1989; H. Rose 1994; Oudshoorn 1994; Clarke 1998, 2005. Also see Ehrenreich and English 1978; Gould 1981; Lewontin 1991.
32 Longino 1990; Rosser 1990; Harding 1993.
33 For more on these feminist scientists, refer to their earlier works, including Benston 1982; Keller 1983; Bleier 1984, 1986; Fausto-Sterling 1985; Birke 1986, 1994; Rosser 1986; Hubbard 1988, 1990, 1995; Haraway 1988; Rogers 1988; Franklin 1990; Messing and Mergler 1995; Spanier 1995; and Barad 1995.
34 van der Tuin 2015.
35 Roy 2004, 2008a.
36 Braidotti 2002, 2006; Grosz 1994, 2011; Kirby 1997; Wilson 1999, 2004; Hird 2004; Alaimo and Hekman 2008; Alaimo 2008; Coole and Frost 2010; Bennett 2010.
37 Benston 1982; Keller 1983.
38 Deleuze and Guattari as discussed in Holland 2013.
39 Deleuze and Guattari as quoted in Adkins 2015, 45.
40 Deleuze and Guattari 1987, 41.
41 Holland 2013, 63.
42 Braidotti 1994, 2002, 2006; Grosz 1993, 1994.
43 Grosz 1993, 176.
44 Grosz 1990.

45 Irigaray 1985.
46 Braidotti 2002, 84.
47 Spivak 1988b, 13–15.
48 Deleuze and Guattari 1987.
49 The "science wars" refers to a series of heated exchanges in the 1990s between scientific realists and postmodernist scholars.
50 Haraway 1988, 1990; Longino 1990, 1993; Harding 1991; Martin 1987; Subramaniam 2000b.
51 Spanier 1995.
52 See chapter 4, "Should Feminists Clone? And If So, How?" During my PhD research, I needed to use the molecular biology technique of subcloning in order to study the possible expression of melatonin and estrogen receptors in gonadotropin-releasing hormone (GnRH) neurons of the hypothalamus. The results of this research is found in Roy and Belsham 2002, and Roy et al. 2001.
53 There is a funny story I must share here. Very soon after I completed my PhD and entered into my first tenure-track position in women's studies, I had the pleasure of meeting Donna Haraway. I was presenting a paper at UC Santa Cruz and although she had not been able to attend my talk, Haraway asked if I would mind sending her my paper. Who does that? With a scholarly generosity that is beyond rare, she not only read my paper within a week's time but also sent me comments in an email. Her comments were detailed, full of new theoretical framings for me to consider. She even provided me with a number of literature sources. I consider this email to be my very first feminist STS love note, and to this day, I do not know why on earth I didn't print it out, frame it, and hang it up in my office. A day after this exchange, our university server crashed, and all of my emails were lost, including the one from Donna-freaking-Haraway. I was too embarrassed to ask her to send me the email again, but I did manage to recall that Haraway was most excited for me to read the work of Isabelle Stengers.
54 Stengers 2000a.
55 Dong and Pierdominici 1995, 25.
56 Harding 2008.
57 Ibid., 215.
58 Colebrook 2002b, xxvi.
59 Harding 2008.
60 Subramaniam 2000b.
61 For Bharati's anticolonial and liberatory thought, see Bharati 1906. Bharati's novel *Jim* (1910–11), which appeared as a serial in *The Light of India* starting in 1907, was written in direct response to the negativity toward India found in Rudyard Kipling's novel *Kim* (1901). In a 2003 Humanities and Social Sciences (HNet) blog post by Gerald Carney, professor of religion at Hampden-Sidney College who was working on a critical study on

Bharati's life and work, Carney states that Bharati wrote *Jim* as a "systematic and virulent refutation of *Kim* in character, plot, culture and ideology" (http://h-net.msu.edu). For anticolonial and liberatory ideas expressed by Tagore, see Tagore 1921 [1916]. Tagore famously renounced his knighthood by the British Crown following the 1919 Jallianwala Bagh Massacre. Lastly, for treatment of J. C. Bose as an anticolonial figure, see Nandy 1995 and Abraham 2006.

62 Nandy 1995, 21.

63 Murphy 2012.

1. Biophilosophies of Becoming

Epigraphs: Bharati 1904, 212; McClintock as quoted in Keller 1983, 200–204; Tagore 1915, 129.

1 Harding 1993, 2008, 2011; Philip 2004; TallBear 2013; Subramaniam 2014.

2 Robinson 2016.

3 Grosz 1994.

4 Keller and Longino 1996; Longino 1990; Sarkar 1996a, 1996b, 2005; Daston and Galison 2007.

5 Said 1979; Mohanty 1984; Spivak 1999; Bhabha 2004.

6 Said 1979; Spivak 1999; Mohanty 1984; Meighoo 2016.

7 Schiebinger 1994; Philip 2004; Harding 2008; TallBear 2013.

8 Lorde 1984a, 116.

9 Ibid., 116; Beauvoir 2011 [1949].

10 Colebrook 2002b, xxxviii.

11 Clayton et al. 2016.

12 I am drawing on the idea of "ontological vulnerability" that has been used by Margrit Shildrick (2004) to discuss the issue of genetics and normativity.

13 Myers 2015, 29.

14 I follow philosopher Keith Ansell Pearson (1999) here, who uses the term "biophilosophy" to describe Deleuze's approach to ethics and matter.

15 Sarkar 1996a, 2005.

16 Sarkar 2005, ix.

17 Sarkar 1996b, 2005.

18 Griffiths 2017.

19 Koutroufinis 2017; 2014, 4, 6.

20 Koutroufinis 2014, 6.

21 Dolphijn and van der Tuin 2012; Grosz 2011; Braidotti 1994.

22 For feminist engagements with process ontology, see Styhre 2001; Haraway 2003; Barad 2007; Braidotti 2010; Grosz 2011; Parisi 2004, 2010; Weinstein 2010; Dolphijn and van der Tuin 2012; van der Tuin 2015. For recent work on process ontology for contemporary biology, see Koutroufinis 2017, 2014; Dupre 2015. For STS engagements with process ontology, see Myers 2015.

23 For recent work on biology and feminism through a philosophy of biology approach, see Hankinson Nelson 2017.
24 Adkins 2015; Colebrook 2002b; Ansell Pearson 1999.
25 Deleuze and Guattari 1987, 261.
26 Ibid.; Deleuze and Guattari 1994.
27 Ansell Pearson 1999, 12.
28 Ibid., 11.
29 Adkins 2015, 1.
30 Penfield 2014.
31 Adkins 2015.
32 Deleuze and Guattari 1987, 20.
33 Adkins 2015, 31 (emphasis in original).
34 Chatterjee 2016, 30.
35 Subramaniam et al. 2017, 410.
36 Ibid.
37 Harding 2016, 1069.
38 Ibid., 1065.
39 Ibid., 1077.
40 Foster 2016, 151.
41 Ibid. Foster draws here from Linda Tuhiwai Smith's (1999) work on decolonizing methodologies.
42 Bignall and Patton 2010, 5; Spivak 1988a.
43 Bignall and Patton 2010, 3.
44 Ibid.
45 Chatterjee 2016, 30.
46 Ibid.
47 Chow 2012, 159.
48 Deleuze and Guattari 1987, 261.
49 Bensmaïa 2017, 21.
50 Ibid., 20.
51 Abraham 2006, 210.
52 Ibid.
53 For more on the postcolonial interventions in STS made by these scholars, refer to Abraham 1998, 2000, 2006; Seth 2009; Anderson 2002, 2009; Harding 2008, 2011; Shiva 1995, 1997; Philip 2004; Philip, Irani, and Dourish 2012; Murphy 2009; Hecht 2002, 2014; Subramaniam 2000a, 2000b, 2014; Roy and Subramaniam 2016; Subramaniam et al. 2017; Foster 2016, 2017; Pollock and Subramaniam 2016; Prasad 2014; Sunder Rajan 2006, 2017.
54 Bharati 1904.
55 For more on this topic, see Subramaniam 2000a; Nanda 2003.
56 Bharati 1904, 8.
57 Ibid., 69.
58 Ibid., 37.
59 Bose 1902.

60 Colebrook 2002b, 109; Bharati 1904, 69.
61 Colebrook 2002b, 65–66.
62 Grosz 2017, 259.
63 McClintock as cited in Keller 1983, 200. For more on the abilities of the nonhuman to scream, see the insightful work on "screaming yeast" by historian of science Sophia Roosth (2009).
64 McClintock 1950.
65 Keller 1983, 200.
66 Ibid.
67 Haraway 1990, 2008, 2015.
68 Haraway 2015, 161.
69 Adkins 2015, 2.
70 Colebrook 2002b, xxv.
71 McClintock as quoted in Keller 1983, 204.
72 Banerji 2015, 4.
73 Barad 2007.
74 Colebrook 2002a, 69.

2. Microphysiologies of Desire

Epigraphs: Bose 1902, 28; Waters and Watson 2015, 2.
1 For a detailed biographical account of Bose's life and his framing as an anticolonial figure, see Nandy 1995 and Abraham 2006.
2 Sen Gupta 2009. In 2014, I had the opportunity to visit the Bose Institute in Kolkata and see these instruments.
3 Ibid.
4 Brahmo Samaj.
5 Sen Gupta, Engineer, and Shepard 2009.
6 Grosz 1993, 171.
7 Bose 1902; Deleuze and Guattari 1994; Grosz 1993; Haraway 2008b; Braidotti 2013.
8 Subramaniam 2014.
9 Ibid., 2.
10 Ibid., 27.
11 Davies, Stacey, and Gilligan 1999.
12 Waters and Watson 2015, 2.
13 Ibid., emphasis added.
14 van der Tuin 2015, 9.
15 Martin 1987.
16 Haraway 1990, 1997; Oudshoorn 1994; Basen, Eichler, and Lippman 1993.
17 Braidotti 2006; Sandoval 2000.
18 McClintock 1950.
19 Rosser 1989; Roy 2004; Roosth and Schrader 2012; Willey and Subramaniam 2017.

20 Roosth and Schrader 2012, 2.
21 Harding 1993; Roberts 1998; Philip 2004; Sunder Rajan 2006; McNeil 2007; Cooper 2008; Hammonds and Herzig 2009; Benjamin 2013; TallBear 2013; Subramaniam 2014; Lee 2013, 2014.
22 Murphy 2009.
23 Latour 2007; Suchman 1987; Wajcman 1991; Haraway 1990.
24 Weber 2006, 400.
25 Ibid.
26 Haraway 1997; Stengers 1997; Barad 2007; Subramaniam 2014; Roy 2004; Kaiser et al. 2009; van Anders 2013; Ritz et al. 2014; Giordano 2016.
27 Weber 2006, 400.
28 Ferrando 2013, 29.
29 Ibid.
30 Haraway 2003.
31 Åsberg 2013, 11, 10.
32 Ibid., 8, emphasis in original.
33 Rosser 1989, 3.
34 Within the same anthology, several authors voiced concern over a separate science that was qualified as a "feminist science." For more on this, see Tuana 1989a, 1989b; Longino 1989; Keller 1989; Harding 1989.
35 Rosser 1989, 3.
36 Harding 1991; Code 1991.
37 Rosser 1989.
38 Ibid., 3.
39 Mills 2010, 146.
40 Ibid.
41 Also see Myers 2015.
42 Roy 2004, 2007, 2008a, 2008b.
43 Mills 2010, 146.
44 Murphy 2012.
45 Boston Women's Health Book Collective 2005.
46 Barad 2007, 19.
47 Ibid., 33, 44, 45.
48 Ibid., 361.
49 Ibid., 175.
50 Buck and Axel 1991.
51 Boehm, Zou, and Buck 2005.
52 Ibid., 683.
53 Zou et al. 2001.
54 Zou et al. 2008, 120.
55 Drug Monkey 2008.
56 Barad 2003, 2007.
57 Haraway 2016, 10.
58 Haraway 2008b, 3.

59 Ibid., 23.
60 Ibid., 24–25.
61 Ibid., 30.
62 Haraway 2016, 12–13.
63 Haraway 2008b, 287.
64 Lather 2007, viii.
65 Ibid., x.
66 Ibid., 10–11.
67 Ibid., 11.
68 Haraway 2016.
69 Lather 2007, 17, 74.
70 Ibid., 11.
71 Ibid., 76, emphasis added.
72 Stengers 2005.
73 Ibid., 195.
74 Stengers 1997.
75 Latour 1997, xiv.
76 Ibid.
77 Shaviro 2005.
78 Kember 2003, 189.
79 Barad 2007, 26.
80 Kember 2003, 189.
81 McClintock as quoted in Keller 1983.
82 Barad 2007, 2012.
83 Bergo 2015; D. Smith 2003.
84 Barad 2007, 364.
85 Ibid., 393.
86 Grosz 1993, 172.
87 Ibid.

3. Bacterial Lives

Epigraphs: Margulis and Sagan 1997, 63; *Gendered Innovations* n.d.
1 Darwin 1859; Margulis 1998.
2 Human Microbiome Project 2017.
3 Sagan 1967.
4 Margulis 1998.
5 Ibid., 73.
6 Margulis 1998; Margulis and Sagan 1986, 1991, 1997, 2002; Sagan and Margulis 1988.
7 Margulis and Sagan 1991, 200.
8 Ibid.
9 Margulis and Sagan 1986, 2.
10 Ibid.

11 Ibid., 3.
12 Margulis and Sagan 1997, 58–59.
13 *Gendered Innovations* n.d.
14 Margulis and Sagan 1991, 201.
15 Stengers 2000a.
16 Epstein 2007.
17 National Science Foundation 2014.
18 National Institutes of Health 2017.
19 US Food and Drug Administration 2013.
20 Ibid.
21 There is an excellent 60 *Minutes* documentary on this topic. See the CBS news story "Sex Matters: 60 *Minutes* Investigates Men, Women and Drug Dose," released on February 9, 2014.
22 Clayton and Collins 2014.
23 Ritz 2016; Joel et al. 2015.
24 Jordan-Young 2010; Fausto-Sterling 2000; Joel 2014.
25 For a history of the scientific research on human sex chromosomes, see Richardson 2013.
26 *Gendered Innovations* n.d.
27 Ibid.
28 Ibid.
29 Ibid.
30 Schiebinger 1999.
31 Collins 1990; Crenshaw 1993.
32 Epstein 2007; Roberts 2012; Pollock 2012; TallBear 2013.
33 An example of the application of this interactionist frame is the UCLA Center for Gender-Based Biology.
34 Oyama, Griffiths, and Gray 2003.
35 Fausto-Sterling 2000, 2005, 2014.
36 Oyama, Griffiths, and Gray 2003, 1.
37 Barad 33, 2007.
38 Haraway 1985.
39 Harding 1986.
40 Rosser 1989.
41 Haraway 1990; Fausto-Sterling 2000; Barad 2003; Subramaniam 2014.
42 Rubin 2012, 891.
43 J. Butler 1999 [1990], 10–11.
44 Ibid., 10.
45 Keller 1985; Harding and O'Barr 1987; Birke 1999; Fausto-Sterling 2000; Kaiser et al. 2009; Jordan-Young 2010.
46 Willey 2016, 124–25.
47 J. Butler 1999 [1990], 11.
48 Ibid.
49 Deleuze and Guattari 1987, 212.

50 Ibid., 213.
51 Grosz 1993.
52 Deleuze and Guattari 1987, 213.
53 Grosz 1994, 14, 18, 181.
54 Parisi 2004, 16–17.
55 Ibid., 187, 61, 4.
56 Ibid., 187.
57 Ibid., 188.
58 Hird 2004, 2009.
59 Hird 2004, 149.
60 Ibid., 151.
61 Hird 2009, 42.
62 Ibid.
63 Ibid., 53.
64 Ibid.
65 Although Jane Bennett (2010) directs her work more toward the nonorganic, she might describe bacteria that talk as a type of vital materialism.
66 Barad 2007, 132.
67 Dolphijn and van der Tuin 2012, 13, 85, 110, 90.
68 van der Tuin 2008, 414; 2011.
69 Dolphijn and van der Tuin 2012; Braidotti 2013, 1994.
70 Haraway 1988.
71 Alaimo 2000, 237; Dolphijn and van der Tuin 2012, 94.
72 Alaimo and Hekman 2008.
73 J. Butler 1993, 30.
74 Kirby 2006; Hekman 2008.
75 Braidotti as cited in Dolphijn and van der Tuin 2012, 27.
76 J. Butler 1986.
77 Dolphijn and van der Tuin 2012, 143.
78 Ibid., 147.
79 J. Butler 1999 [1990], 11; Kirby 2008, 214 and 219.
80 Breen et al. 2001; Kirby 2008, 2011, 2012.
81 Kirby 2008, 219.
82 Kirby as quoted in Breen et al. 2001, 13.
83 J. Butler as quoted in ibid.
84 Kirby 2008, 219.
85 Kirby 2012, 200.
86 Bugg et al. 2011.
87 Tavares et al. 2013; Lambert 2005; Lawrence 1999.
88 Kirby 2008, 2011, 2012.
89 Kirby 1997, 60.
90 Ibid., 60–61.
91 Ibid., 61, emphasis in original.

92 Kirby 2006, 84.
93 Haraway 2003; Braidotti 2013.
94 Kirby 2012, 200.
95 Tuana 2008, 209.
96 Alaimo and Hekman 2008, 7–8.
97 Kirby 2006, 84.
98 Kirby 2008, 219.
99 Butler as quoted in Breen et al. 2001, 13.
100 Alaimo and Hekman 2008, 8.
101 Although I find the insertion of human-created poetry into bacterial genomes, as in Christian Bök's work in *The Xenotext: Book 1* (2015), to be of interest, I would align this project with other biotechnologies that have turned to bacteria for their mechanistic appeal. The bacterial poetry that I am thinking about here is writing that bacteria conducts that we as humans might learn to appreciate outside of our typical productionist frameworks.
102 Haraway 2008a, 178.
103 Derrida 1997, 158; Kirby 1997, 60.
104 Derrida 1988, 136 and 152.
105 Philip 2004; McNeil 2005; Reardon 2005; Sunder Rajan 2006; Cooper 2008; Harding 2008; Murphy 2012; Pollock 2012; TallBear 2013; Benjamin 2013; Roy and Subramaniam 2016; Subramaniam et al. 2017; Foster 2017.
106 Sunder Rajan 2006, 3, 6; emphases in original.
107 Cooper 2008, 4.
108 Ibid., 33.
109 Ibid., 9.
110 Philip 2004, 197.

4. Should Feminists Clone?

Epigraphs: Braidotti 2002, 10; Massumi 1987, 95.

1 Deleuze and Guattari 1987, 272, 11.
2 Ibid., 276.
3 In Bruno Latour's foreword to Stengers's *Power and Invention* (1997), he discusses Stengers's notion of "risk" as being integral to properly constructed propositions. According to Stengers, although constructions without risk may be moral and politically correct, they may not be CC (cosmopolitically correct) (Latour 1997, xiv). Although my effort may not fully measure up, the risk I run by posing the question "Should feminists clone?" reflects my goal to be CC.
4 Cloning is often used as shorthand for the molecular biology technique of subcloning.
5 The Biology As If the World Matters (BAITWorM) panel was held at the Ontario Institute for Studies in Education, University of Toronto. The

BAITWorM Network consisted of natural and social scientists committed to interdisciplinary research and teaching in the sciences.

6 Corea 1985.

7 The phrase "technical, organic, and political" is commonly used by Haraway (1991; 1997). The technical, organic, and political also comprise three of the six imploding categories that inform her practice of figuration (Haraway 1997).

8 Braidotti 2006, 12.

9 Massumi 1987, 90.

10 Ibid., 96.

11 Ibid., 96–97.

12 Baudrillard 1983, 4.

13 Massumi 1987, 96.

14 Ibid., 97.

15 Ibid., 91.

16 Parr 2005, 223.

17 Wilson 1999, 8.

18 Keane and Rosengarten 2002, 261.

19 Wilson 1999, 8.

20 Haraway 2004, 9.

21 Rouse 1996, 215–19.

22 Braidotti 2002, 169.

23 Stengers 2000b, 155.

24 New England Biolabs 2018.

25 ThermoFisher Scientific 2017.

26 Ibid., 4.

27 Haraway 1997.

28 Ibid., 3.

29 In her elegant fictional story "Confessions of a Bioterrorist," Charis Thompson (1999) writes about a reproductive biologist who yearns to reproduce. In this case, however, the resident biologist for the San Diego Zoo yearns to give birth to human-bonobo interspecies offspring.

30 Haraway 1998; Sandoval 2000; McClintock as cited in Keller 1983.

31 Sandoval 2000, 104, 83.

32 Martin 1987.

33 Roy, Angelini, and Belsham 1999.

34 Haraway (1997, 12) has articulated six categories that "inform the practice of figuration," including the "technical, organic, political, economic, oneiric and textual."

35 Diprose and Ferrell 1991, ix.

36 Wilson 1998, 201.

37 Deleuze and Guattari 1987, 12.

38 Sandoval 2000, 114.

39 Haraway 1997, 129.

40 This is a derogatory and racist term used to describe an individual of South Asian descent, derived from the term "Pakistani."
41 Haraway 1997, 129.
42 Roy, Angelini, and Belsham 1999.
43 ThermoFisher Scientific 2017.
44 Keller 1983, 198.
45 Ibid., 8.
46 Ibid., 194.
47 Diprose and Ferrell 1991, ix.
48 ThermoFisher Scientific 2017.
49 Firestone 2003, 29.
50 Ibid., 180.
51 Ibid., 164.
52 Ibid., 183.
53 Ibid., 185, emphasis added.
54 The revisioning of society, where only women reside and reproduction occurs through parthenogenesis, can be found in such novels as Charlotte Perkins Gilman's *Herland* (1979 [1915]), Joanna Russ's *The Female Man* (1975), and Suzy McKee Charnas's *Motherlines* (1978). Technologies such as cloning are used for reproduction in Pamela Sargent's *Cloned Lives* (1978), Kate Wilhelm's *Where Late the Sweet Birds Sang* (1998 [1976]), and Ursula Le Guin's short story "Nine Lives" (1976 [1969]). Some other form of genetic manipulation occurs for the purpose of reproduction in Marge Piercy's *Women on the Edge of Time* (1976) and Octavia Butler's *Dawn* (1987).
55 Sargent 1976, xix.
56 I was unable to find out the contraceptive for which the clinical trial was being conducted, or if it was a contraceptive at all, but the patches looked very similar to the Ortho Evra patch. A local woman informed me that they were participating in a government-subsidized family-planning incentive program. By participating in the trial, these women were provided with food rations and shoes for their children.
57 Bhatia 2005.
58 Times of India 2016.
59 Kfoury 2007, 112.
60 For a more detailed analysis of stem cell research in California, see Benjamin 2013.
61 Murugan 2009.
62 In this announcement Bush limited public funding of human embryonic stem cell research to seventy-eight cell lines that had already been established before August 9, 2001. It was later reported that of these seventy-eight cell lines, only nineteen were actually viable (*USA Today* 2005).
63 MSNBC Online 2004.
64 Adelson and Weinberg 2010.
65 *USA Today* 2007.

66 NIH Stem Cell Information Home Page 2016.
67 NBC News 2017.
68 Hiltzik 2017.
69 Waldby 2002; Waldby and Carroll 2012.
70 Salleh 2007.
71 BBC News 2005.
72 NPR 2006.
73 ViaGen 2017.
74 Sandoval 2000.
75 Haraway 1991; Sandoval 2000.
76 Haraway 1991, 150.
77 Sandoval 2000, 167.
78 Ibid.
79 Ibid., 30.
80 Ibid.
81 Roy, Angelini, and Belsham 1999.
82 Stengers 2000a, 41.
83 Stengers 2005, 187.
84 Ibid., 187–88.
85 Ibid., 188. Stengers explains here that her ecology of practices is composed of what Brian Massumi has referred to as a Leibnizian technology.
86 Braidotti 2006, 15.
87 Ibid., 83. In her discussion of poststructural ethics, Braidotti (ibid.) states that becoming political involves "a radical repositioning or internal transformation on the part of subjects who want to become-minoritarian in a productive and affirmative manner."
88 Ibid., 134.

5. *In Vitro* Incubations

Epigraphs: Barad 2007, 172; Ansell Pearson 1999, 20.
1 Landecker 2007, 7.
2 Ibid., 26.
3 Ibid., 3.
4 Wetsel et al. 1991.
5 RU486, also referred to as mifepristone, is an antiprogestin used for medical abortions. In the presence of progesterone, RU486 acts as a competitive progesterone antagonist (a ligand or drug that blocks a biological response). However, in the absence of progesterone, it acts as a partial agonist (a ligand or drug that binds to a receptor and activates a biological response). For reviews on the molecular mechanisms of RU486, see Kakade and Kulkarni 2014; Mahajan and London 1997.
6 Alaimo and Hekman 2008; Coole and Frost 2010.
7 Thacker 2010, 220.

8 Haraway 2003; Barad 2007; Bennett 2010; Chen 2012.
9 Foucault 1970, 160–61.
10 Deleuze and Guattari 1987, 499, emphasis in original.
11 Haraway 2015, 116.
12 Grosz 2008, 24.
13 Kirby 2006, 84.
14 Barad 2010, 264–65.
15 Jagger 2015, 336.
16 Landschoot 2018.
17 Thanks to a grant I received in 2010 from the National Academies KECK Futures Initiative, I was able carry out my project "Developing a Bench-side Ethics and Community Based Participatory Research Training Program in Synthetic Biology." Dr. Sara Giordano was the postdoctoral fellow on this project and has recently published an article on synthetic biology, feminist pedagogy, and bioethics (Giordano 2016).
18 Thacker 2010, ix.
19 Ibid., xiii.
20 Ibid.
21 Ibid., 141. Thacker refers here to Deleuze's ontology in *Difference and Repetition* (1994).
22 Haraway 1997.
23 Thacker 2010, 219, emphasis in original.
24 Ibid.
25 Ansell Pearson 1999, 4; Dema 2007.
26 Ansell Pearson 1999, 152.
27 See Dema 2007. John Protevi (2012, 239) has explained that we should "remember Deleuze's love of provocation" and his "idiosyncratic notion of vitalism" when we consider how vitalism operates in Deleuzian thinking.
28 Protevi 2012, 247.
29 Deleuze and Guattari 1987, 15, 158.
30 Ansell Pearson 1999, 154.
31 Dema 2007.
32 Ibid.
33 Deleuze and Guattari 1987, 22.
34 Bryant 2011b.
35 Ansell Pearson 1999, 141–42.
36 Ibid., 2.
37 Deleuze and Guattari 1987, 4.
38 Protevi 2012, 248.
39 Barad 2010, 264.
40 Foucault 1970, xii–xiii.
41 Describing Deleuze and Guattari's three strata, Eugene Holland (2013, 23) writes that "the inorganic stratum starts with a bang—with the Big Bang, that is."

42 Protevi 2012, 251.

43 For a more extensive treatment of the role of central dogmas in the history and philosophy of molecular biology, see Sarkar 1996a, 1996b, and 2005.

44 Breen et al. 2001, 13.

45 For rich historical accounts of the people, funding sources, and universities involved in the development of molecular biology and genetics research in the United States, see the work of Lily E. Kay (1993, 1996, 2000).

46 Crick 1988, 109.

47 Protevi 2012, 251.

48 In the PBS television series *DNA: Pandora's Box* (2003), James Watson speaks about the importance of eugenics and how the events of Nazi Germany have given eugenics a bad name (Duncan and Glover 2003).

49 BioBricks Foundation 2017.

50 Lentzos et al. 2008; Fritz et al. 2010; Ausländer, Ausländer, and Fussenegger 2017.

51 Endy 2008, 1196–97.

52 Hutchison et al. 2016. The first synthetic minimal yeast genome has also recently been reported. This minimal genome is a reduced version of the genome derived from the yeast species *Saccharomyces cerevisiae*. See Richardson et al. 2017.

53 Lentzos et al. 2008, 316.

54 Rabinow and Bennett 2009.

55 Protevi 2012, 251.

56 N. Rose 2007, 3.

57 Ibid., 4.

58 Ibid.

59 Protevi 2012, 251.

60 Eimer 2010, 75, emphasis in original.

61 Hutchison et al. 2016.

62 Glass et al. 2017.

63 Lartigue et al. 2007.

64 Hutchison et al. 2016.

65 Engineering and Physics Research Council 2017.

66 Ibid.

67 Roosth 2009.

68 Ibid., 337.

69 Holland 2013, 24.

70 Shou, Ram, and Vilar 2007.

71 Dunham 2007, 1741.

72 Shou, Ram, and Vilar. 2007, 1877.

73 Ibid.

74 Ibid.

75 Breen et al. 2001, 13.

76 Cooper 2008.

77 Holland 2013, 29.
78 Chatterjee 2016, 30.
79 Deleuze and Guattari 1987, 60.
80 Welchman 1997, 224; Bryant 2011a; Harding 2016; Foster 2016.
81 Gibson et al. 2008.
82 Hutchison et al. 2016.
83 ETC Group 2007, 2016.
84 J. Craig Venter's full bio is available on the JCVI website, at www.jcvi.org/cms/about/bios/jcventer.
85 Venter used the shotgun technique once again when he set out to beat the publicly funded sequencing effort of the Human Genome Project in the mid-1990s.
86 Glass et al. 2006, 425.
87 Sweet 2009, 823.
88 Hartmann 2009, 371.
89 Cooper 2008, 58.
90 Keasling 2009.
91 Farooq and Mahajan 2004. For several years now the World Health Organization (2015) has recommended the use of artemisinin-based combination therapies for the treatment of malaria, a disease caused by a plasmodium parasite that infects red blood cells and leads to severe disease and death.
92 Ro et al. 2006, 940.
93 Ibid.; Westfall et al. 2012.
94 Paddon et al. 2013, 528.
95 The pharmaceutical giant Sanofi Aventis took on the manufacturing of semisynthetic artemisinin. The product hit the market in 2016 apparently with a fair bit of resistance. Without the proper market demand, semisynthetic artemisinin is no longer being touted as a panacea for malaria (see Peplow 2016).
96 Sandoval et al. 2014, 215–16.
97 MacKenzie 2013, 194.
98 Ibid., 190.
99 Edel 1969; Ebert 2008.
100 Ortiz 1947, 268.
101 Ibid., 98, 101.
102 Amyris recently announced that it will build a 23,000-ton production plant in Queensland, Australia, to produce farnesene for use in cosmetics. Sugarcane will serve as the feedstock in their operations and with the proximity to Asia, the hope is to tap into the Asian cosmetics market (see biofuelsdigest.com).
103 Mohanty 2003, 229.
104 This sentiment can be found in Thomas's eloquent public debate with Drew Endy and is an excellent example of postcolonial and decolonial STS practices at work. For more, see the Long Now Foundation 2008.

105 Venter et al. 2004, 66.

106 Although the website is no longer active, Venter's voyages aboard the *Sorcerer II* were recorded in detail at www.sorcerer2expedition.org. As one entered into Venter's virtual world aboard the website for his expedition vessel, a map of the planet and the path of scientific progress could be seen as it is rolled out to the sound of splashing waves. The information available on the website was divided into sections titled "Expedition Info," "Voyage Tracker," "Sampling Methods," and "Scientific Data."

107 Ajjawi et al. 2017.

108 Ibid., 652.

109 See Synthetic Genomics 2017.

110 In 2017, ExxonMobil released the commercial "Energy Farmer," which references the development of biofuels.

111 According the JCVI website, Synthetic Genomics Inc., the private company held by J. Craig Venter, holds intellectual property rights to the tools and technologies developed for the synthetic engineering of organisms and has filed thirteen patent family applications. Patents approved as of 2017 include digital-to-biological converter, programmable oligonucleotide synthesis, assembly of large nuclei acids, method for producing polymers, *in vitro* recombination method, and more (see http://patents.justia.com/assignee/synthetic-genomics-inc).

112 Duncan and Selim 2004.

113 Russell 2004.

114 This episode of *The Colbert Show* with Venter aired on October 20, 2007, on Comedy Central.

115 Spivak 1985, 243, 253, 243.

116 Ibid., 241.

117 Ibid., 252.

118 Ibid., 253–54.

119 Ibid., 240, 241.

Conclusion

Epigraphs: *Oxford English Dictionary*, s.v. "grass root," accessed January 14, 2018, http://www.oed.com.proxy.library.emory.edu/view/Entry/80912?redirectedFrom=grassroots#eid. "Grass roots" is from *Oxford Living Dictionaries*, accessed January 14, 2018. https://en.oxforddictionaries.com/definition/grass_roots.

1 Black Lives Matter 2018.

2 Minnite and Piven 2016, 275.

3 Ibid., 278.

4 Ibid., 277.

5 Black Lives Matter 2018.

Bibliography

Abraham, Itty. 1998. *The Making of the Indian Atomic Bomb*. New York: Zed Books.

———. 2000. "Landscape and Postcolonial Science." *Contributions to Indian Sociology* 34(2):163–87.

———. 2006. "The Contradictory Spaces of Postcolonial Technoscience." *Economic and Political Weekly* 41(3):210–17.

Adelson, Joel, and Joanna Weinberg. 2010. "The California Stem Cell Initiative: Persuasion, Politics, and Public Science." *American Journal of Public Health* 100(3):446–51.

Adkins, Brent. 2015. *Deleuze and Guattari's A Thousand Plateaus: A Critical Introduction and Guide*. Edinburgh: Edinburgh University Press.

Ajjawi, Imad, et al. 2017. "Lipid Production in *Nannochloropsis gaditana* Is Doubled by Decreasing Expression of a Single Transcriptional Regulator." *Nature Biotechnology* 35(7):647–52.

Alaimo, Stacy. 2000. *Undomesticated Ground: Recasting Nature as Feminist Space*. Ithaca, NY: Cornell University Press.

———. 2008. "Transcorporeal Feminisms and the Ethical Space of Nature." In *Material Feminisms*, ed. Stacy Alaimo and Susan Hekman, 237–64. Bloomington: Indiana University Press.

Alaimo, Stacy, and Susan Hekman. 2008. "Introduction: Emerging Models of Materiality in Feminist Theory." In *Material Feminisms*, ed. Stacy Alaimo and Susan Hekman, 1–19. Bloomington: Indiana University Press.

Alcoff, Linda. 2006. *Visible Identities: Race, Gender and the Self*. New York: Oxford University Press.

Anderson, Warwick. 2002. "Introduction: Postcolonial Technoscience." *Social Studies of Science* 32(5/6):643–58.

———. 2009. "From Subjugated Knowledge to Conjugated Subject: Science and Globalization, or Postcolonial Studies of Science?" *Postcolonial Studies* 12(4):389–400.

Ansell Pearson, Keith. 1999. *Germinal Life: The Difference and Repetition of Deleuze*. New York: Routledge.

Åsberg, Cecilia. 2013. "The Timely Ethics of Posthumanist Gender Studies." *feministische studien* 1:7–12.
Ausländer, Simon, David Ausländer, and Martin Fussenegger. 2017. "Synthetic Biology—The Synthesis of Biology." *Angewandte Chemie International Edition* 56:6396–419.
Baker, Lauren, Kirstan Meldrum, Meijing Wang, Rajakumar Sankula, Ram Vanam, Azad Raiesdana, Ben Tsai, Karen Hile, John Brown, and Daniel Meldrum. 2003. "The Role of Estrogen in Cardiovascular Disease." *Journal of Surgical Research* 115(2):325–44.
Banerji, Debashish. 2015. "Theory and the Performative Politics of Punctuation." In *Rabindranath Tagore in the 21st Century: Theoretical Renewals*, ed. Debashish Banerji, 1–14. New York: Springer.
Barad, Karen. 1995. "A Feminist Approach to Teaching Quantum Physics." In *Teaching the Majority: Breaking the Gender Barrier in Science, Mathematics, and Engineering*, ed. Sue V. Rosser, 79–97. New York: Teachers College Press.
———. 2003. "Posthumanist Performativity: Toward an Understanding of How Matter Comes to Matter." *Signs: Journal of Women in Culture and Society* 28(3):801–31.
———. 2007. *Meeting the Universe Halfway: Quantum Physics and the Entanglement of Matter and Meaning*. Durham, NC: Duke University Press.
———. 2010. "Quantum Entanglements and Hauntological Relations of Inheritance: Dis/continuities, SpaceTime Enfoldings, and Justice-to-Come." *Derrida Today* 3(2):240–68.
———. 2012. "On Touching—The Inhuman That Therefore I Am." *differences* 23(3):206–23.
Barakat, Radwa, Oliver Oakley, Heehyen Kim, Jooyoung Jin, and CheMyong Jay Ko. 2016. "Extra-gonadal Sites of Estrogen Biosynthesis and Function." *Biochemistry and Molecular Biology Reports* 49(9):488–96.
Basen, Gwynne, Margrit Eichler, and Abby Lippman, eds. 1993. *Misconceptions: The Social Construction of Choice and the New Reproductive and Genetic Technologies*. Volume 1, 2. Hull, Quebec, Canada: Voyageur Publishing.
Baudrillard, Jean. 1983. *Simulations*. New York: Semiotext(e).
BBC News. 2005. "UN Vote Urges Human Cloning Ban." March 8. http://news.bbc.co.uk/1/hi/health/4328919.stm.
Beauvoir, Simone de. 2011 [1949]. *The Second Sex*. Translated by Constance Borde and Sheila Malovany-Chevallier. New York: Vintage Books.
Belsham, Denise D., and David Lovejoy. 2005. "Gonadotropin-releasing Hormone: Gene Evolution, Expression and Regulation." *Vitamins and Hormones* 71:59–94.
Bennett, Jane. 2010. *Vibrant Matter: A Political Ecology of Things*. Durham, NC: Duke University Press.
Benjamin, Ruha. 2013. *People's Science: Bodies and Rights on the Stem Cell Frontier*. Stanford, CA: Stanford University Press.

Bensmaïa, Réda. 2017. *Gilles Deleuze, Postcolonial Theory, and the Philosophy of Limit*. New York: Bloomsbury.
Benston, Margaret. 1982. "Feminism and the Critique of the Scientific Method." In *Feminism in Canada*, ed. Angela Miles and Geraldine Finn, 47–66. Montreal: Black Rose.
Bergo, Bettina. 2011 "Emmanuel Levinas." *The Stanford Encyclopedia of Philosophy*, ed. Edward N. Zalta. https://plato.stanford.edu/archives/sum2015/entries/levinas.
Bhabha, Homi. 2004. *The Location of Culture*. New York: Routledge Classics.
Bharati, Baba Premananda. 1904. *Sree Krishna: The Lord of Love*. New York: S.L. Parsons and Company.
———. 1906. "The White Peril." *The Light of India* 1(2):41–47.
Bhatia, Rajani. 2005. "Ten Years after Cairo: The Resurgence of Coercive Population Control in India." *Different Takes: A Publication of the Population and Development Program, Hampshire College* 31. http://popdev.hampshire.edu/projects/dt/dt31.php.
Bignall, Simone, and Paul Patton. 2010. "Introduction: Deleuze and the Postcolonial: Conversations, Negotiations, Mediations." In *Deleuze and the Postcolonial*, ed. Simone Bignall and Paul Patton, 1–19. Edinburgh: Edinburgh University Press.
BioBricks Foundation. 2017. Home page. Accessed July 25, 2018. https://biobricks.org.
Biofuels Digest. 2017. "Two Up: Amyris, Queensland Gamble on Farnesene for SE Asia." Accessed August 30, 2018. www.biofuelsdigest.com/bdigest/2017/06/22/two-up-amyris-queensland-gamble-on-farnesene-for-se-asia.
Birke, Lynda. 1986. *Women, Feminism and Biology: The Feminist Challenge*. Upper Saddle River, NJ: Prentice Hall/Harvester Wheatsheaf.
———. 1994. *Feminism, Animals and Science: The Naming of the Shrew*. Bristol: Open University Press.
———. 1999. *Feminism and the Biological Body*. New Brunswick, NJ: Rutgers University Press.
Black Lives Matter. 2018. "What We Believe." Accessed January 14, 2018. http://blacklivesmatter.com/about/what-we-believe.
Bleier, Ruth. 1984. *Science and Gender: A Critique of Biology and Its Theories on Women*. New York: Pergamon Press.
———. 1986. *Feminist Approaches to Science*. New York: Pergamon Press.
Boehm, Ulrich, Zhihua Zou, and Linda B. Buck. 2005. "Feedback Loops Link Odor and Pheromone Signaling with Reproduction." *Cell* 123(4):683–95.
Bök, Christian. 2015. *Xenotext: Book 1*. Toronto: Coach House Books.
Bose, Jagadish Chandra. 1902. *Response in the Living and Non-Living*, ed. Perfect Library. Lexington, KY: Create Space Independent Publishing Platform.
Boston Women's Health Book Collective. 2005. *Our Bodies Ourselves*. New York: Touchstone.

Brahmo Samaj. n.d. "India's National Religion." Accessed June 24, 2017. www.brahmosamaj.in.
Braidotti, Rosi. 1994. *Nomadic Subjects: Embodiment and Sexual Difference in Contemporary Feminist Theory*. New York: Columbia University Press.
———. 2002. *Metamorphosis: Towards a Materialist Theory of Becoming*. Cambridge, MA: Polity Press.
———. 2006. *Transpositions: On Nomadic Ethics*. Cambridge, MA: Polity Press.
———. 2010. "On Putting the Active Back into Activism." *New Formations: A Journal of Culture/Theory/Politics* 68:26–57.
———. 2013. *The Posthuman*. Cambridge, MA: Polity Press.
Breen, Margaret Soenser, Warren J. Blumenfeld, Susanna Baer, Robert A. Brookey, Lynda Hal, Vicky Kirby, Diane Helene Miller, Robert Shail, and Natalie Wilson. 2001. "'There Is a Person Here': An Interview with Judith Butler." *International Journal of Sexuality and Gender Studies* 6(1/2):7–23.
Bryant, Levi R. 2011a. "Deterritorialization." July 2. https://larvalsubjects.wordpress.com/2011/07/02/deterritorialization.
———. 2011b. "Two Types of Assemblages." February 20. https://larvalsubjects.wordpress.com/2011/02/20/two-types-of-assemblages.
Buck, Linda, and Richard Axel. 1991. "A Novel Multigene Family May Encode Odorant Receptors: A Molecular Basis for Odor Recognition." *Cell* 65(1):175–87.
Bugg, T.D.H., D. Braddick, C. G. Downson, and D. I. Roper. 2011. "Bacterial Cell Wall Assembly: Still an Attractive Antibacterial Target." *Trends in Biotechnology* 29(4):167–73.
Butler, Judith. 1986. "Sex and Gender in Simone de Beauvoir's Second Sex." *Yale French Studies* 72:35–49.
———. 1993. *Bodies That Matter: On the Discursive Limits of "Sex."* New York: Routledge.
———. 1999 [1990]. *Gender Trouble: Feminism and the Subversion of Identity*. New York: Routledge.
Butler, Octavia. 1987. *Dawn*. New York: Warner Books.
Charnas, Suzy McKee. 1978. *Motherlines*. New York: Putnam Books.
Chatterjee, Sushmita. 2016. "What Does It Mean To Be a Postcolonial Feminist? The Artwork of Mithu Sen." *Hypatia* 31(1):22–40.
Chen, Mel Y. 2012. *Animacies: Biopolitics, Racial Mattering, and Queer Affect*. Durham, NC: Duke University Press.
Chow, Rey. 2012. *Entanglements, or Transmedial Thinking about Capture*. Durham, NC: Duke University Press.
Clarke, Adele. 1998. *Disciplining Reproduction: Modernity, American Life Sciences, and the Problems of Sex*. Berkeley: University of California Press.
———. 2005. *Situational Analysis: Grounded Theory after the Postmodern Turn*. Thousand Oaks, CA: Sage Publications.
Clayton, Janine A., and Francis S. Collins. 2014. "Policy: NIH to Balance Sex in Cell and Animal Studies." *Nature* 509:282–83.

Clayton, W. D., M. S. Vorontsova, K. T. Harmon, and H. Williamson. 2016. "GrassBase—The Online World of Grass Flora." January 20. www.kew.org/data/grassbase/news.htm.

Code, Lorraine. 1991. *What Can She Know? Feminist Theory and the Construction of Knowledge*. Ithaca, NY: Cornell University Press.

Colebrook, Claire. 2002a. *Gilles Deleuze: Essential Guides for Literary Studies*. New York: Routledge.

———. 2002b. *Understanding Deleuze*. Crows Nest, New South Wales, Australia: Allen and Unwin.

Collins, Patricia Hill. 1990. *Black Feminist Thought: Knowledge, Consciousness, and the Politics of Empowerment*. New York: Routledge.

Coole, Diana, and Samantha Frost. 2010. *New Materialisms: Ontology, Agency, and Politics*. Durham, NC: Duke University Press.

Cooper, Melinda. 2008. *Life as Surplus: Biotechnology and Capitalism in the Neoliberal Era*. Seattle: University of Washington Press.

Corea, Gena. 1985. *The Mother Machine: Reproductive Technologies from Artificial Insemination to Artificial Wombs*. New York: Harper Collins.

Crenshaw, Kimberlé. 1993. "Mapping the Margins: Intersectionality, Identity Politics, and Violence against Women of Color." *Stanford Law Review* 43:1241–99.

Crick, Francis. 1988. *What Mad Pursuit: A Personal View of Scientific Discovery*. New York: Basic Books.

Darwin, Charles. 1859. "On the Origin of Species by Means of Natural Selection, or the Preservation of Favored Races in the Struggle for Life." In *So Simple a Beginning: Darwin's Four Great Books*, ed. E. O. Wilson (2005). New York: W.W. Norton and Company.

Daston, Lorraine, and Peter Galison. 2007. *Objectivity*. Brooklyn: Zone Books.

Davies, Julia M., Adrian J. Stacey, and Christopher A. Gilligan. 1999. "*Candida albicans* Hyphal Invasion: Thigmotropism or Chemotropism?" *FEMS Microbiology Letters* 171:245–49.

Deleuze, Gilles. 1994. *Difference and Repetition*. Translated by Paul Patton. Originally published in 1968. New York: Columbia University Press.

Deleuze, Gilles, and Félix Guattari. 1987. *A Thousand Plateaus: Capitalism and Schizophrenia*. Translated by Brian Massumi. Minneapolis: University of Minnesota Press.

Dema, Leslie. 2007. "'Inorganic, Yet Alive': How Can Deleuze and Guattari Deal with the Accusation of Vitalism?" *Rhizomes: Cultural Studies in Emerging Knowledge* 15 (Winter). www.rhizomes.net/issue15/dema.html#_ftnref1

Derrida, Jacques. 1988. *Limited Inc*. Evanston, IL: Northwestern University Press.

———. 1997. *Of Grammatology*. Translated by Gayatri Chakravorty Spivak. Baltimore, MD: John Hopkins University Press.

Diprose, Rosalyn, and Robyn Ferrell. 1991. Introduction to *Cartographies: Poststructuralism and the Mapping of Bodies and Spaces*, ed. Rosalyn Diprose and Robyn Ferrell, viii–xi. North Sydney: Allen and Unwin.

Dolphijn, Rick, and Iris van der Tuin. 2012. *New Materialism: Interviews and Cartographies*. Ann Arbor, MI: Open Humanities Press.
Dong, Ming, and Maria Grazia Pierdominici. 1995. "Morphology and Growth of Stolons and Rhizomes in Three Clonal Grasses, As Affected by Different Light Supply." *Vegetatio* 116:25–32.
Drug Monkey. 2008. "How to Read a Retraction." March 5. http://drugmonkey.scientopia.org/2008/03/05/how-to-read-a-retraction.
Duncan, David Ewing, and Joycelyn Selim. 2004. "Discover Dialogue: Geneticist Craig Venter." *Discover*. December 3. http://discovermagazine.com/2004/dec/discover-dialogue.
Duncan, Ian, and David Glover. 2003. *DNA: Pandora's Box*. PBS Television Series.
Dunham, Maitreya J. 2007. "Synthetic Ecology: A Model System for Cooperation." *Proceedings of the National Academy of Sciences USA* 104(6):1741–42.
Dupre, John. 2015. "A Process Ontology for Biology." *Physiology News* 100:33–34.
Ebert, Christopher. 2008. *Between Empires: Brazilian Sugar in the Early Atlantic Economy, 1550–1630*. Leiden: Brill Publishers.
Edel, Matthew. 1969. "The Brazilian Sugar Cycle of the Seventeenth Century and the Rise of West Indian Competition." *Caribbean Studies* 9(1):24–44.
Ehrenreich, Barbara, and Deidre English. 1978. *For Her Own Good: 150 Years of the Experts' Advice to Women*. New York: Anchor Books.
Eimer, Sara Dawn. 2010. "Love, Desire, Individuation: Intersections of Plato, Schelling and Deleuze." PhD dissertation. Pennsylvania State University.
Endy, Drew. 2008. "Reconstruction of the Genomes." *Science* 319:1196–97.
Engineering and Physics Research Council. 2017. "Towards a Universal Biological-Cell Operating System (AUdACiOuS)." http://gow.epsrc.ac.uk/NGBOViewGrant.aspx?GrantRef=EP/J004111/1.
Epstein, Steve. 2007. *Inclusion: The Politics of Difference in Medical Research*. Chicago: University of Chicago Press.
ETC Group. 2007. "Synthia's Last Hurdle?" June 7. www.etcgroup.org/content/synthia%E2%80%99s-last-hurdle.
———. 2016. "Craig Venter Lays an Easter Egg." March 24. www.etcgroup.org/content/craig-venter-lays-easter-egg.
Farooq, Umar, and R. C. Mahajan. 2004. "Drug Resistance in Malaria." *Journal of Vector Borne Diseases* 41:45–53.
Fausto-Sterling, Anne. 1985. *Myths of Gender: Biological Theories about Women and Men*. New York: Basic Books.
———. 2000. *Sexing the Body: Gender Politics and the Construction of Sexuality*. New York: Basic Books.
———. 2005. "The Bare Bones of Sex: Part I—Sex and Gender." *Signs: Journal of Women in Culture and Society* 30(2):1491–1527.
———. 2014. "How Your Generic Baby Acquires Gender." Podcast. http://www3.unil.ch/wpmu/neurogenderings3/podcasts.
Ferrando, Francesca. 2013. "Posthumanism, Transhumanism, Antihumanism, Metahumanism, and New Materialisms: Differences and Relations."

Existenz: An International Journal in Philosophy, Religion, Politics, and the Arts 8(2):26–32.
Firestone, Shulamith. 2003 [1970]. *The Dialectic of Sex: The Case for Feminist Revolution*. New York: Farrar, Straus and Giroux.
Foster, Laura. 2016. "Decolonizing Patent Law: Postcolonial Technoscience and Indigenous Knowledge in South Africa." *Feminist Formations* 28(3):148–73.
———. 2017. *Reinventing Hoodia: Peoples, Plants, and Patents in South Africa*. Seattle: University of Washington Press.
Foucault, Michel. 1970. *The Order of Things: An Archeology of the Human Sciences*. New York: Vintage.
Franklin, Ursula. 1990. *The Real World of Technology*. Toronto: Anansi.
Fritz, Brian R., Laura E. Timmerman, Nichole M. Daringer, Joshua N. Leonard, and Michael C. Jewett. 2010. "Biology by Design: From Top to Bottom and Back." *Journal of Biomedicine and Biotechnology* (e-publication):1–11. Article ID 232016.
Gendered Innovations in Science, Health and Medicine, Engineering, and Environment. n.d. "Terms." Accessed July 17, 2017. http://genderedinnovations.stanford.edu/terms.html.
Gibson, D. G., et al. 2008. "Complete Chemical Synthesis, Assembly, and Cloning of a Mycoplasma Genitalium Genome." *Science* 319(5867):1215–20.
Gilman, Charlotte Perkins. 1979 [1915]. *Herland*. New York: Pantheon Books.
Giordano, Sara. 2016. "Building New Bioethical Practices through Feminist Pedagogies." *International Journal of Feminist Approaches to Bioethics* 9(1):81–103.
Glass, John I., Chuck Merryman, Kim S. Wise, Clyde A. Hutchison, and Hamilton O. Smith. 2017. "Minimal Cells—Real and Imagined." *Cold Spring Harbor Perspectives in Biology.* doi: 10.1101/cshperspect.a023861.
Glass, John I., Nacyra Assad-Garcia, Nina Alperovich, Shibu Yooseph, Matthew R. Lewis, Mahir Maruf, Clyde A. Hutchison III, Hamilton O. Smith, and J. Craig Venter. 2006. "Essential Genes of a Minimal Bacterium." *Proceedings of the National Academy of Sciences* 103(2):425–30.
Gould, Stephen J. 1981. *The Mismeasure of Man*. New York: Norton.
Griffiths, Paul. 2017. "Philosophy of Biology." *The Stanford Encyclopedia of Philosophy*, ed. Edward Zalta. https://plato.stanford.edu/entries/biology-philosophy.
Grosz, Elizabeth. 1990. "Conclusion: A Note on Essentialism and Difference." In *Feminist Knowledge: Critique and Construct*, ed. Sneja Gunew, 332–44. New York: Routledge.
———. 1993. "A Thousand Tiny Sexes: Feminism and Rhizomatics." *Topoi* 12:167–79.
———. 1994. *Volatile Bodies: Towards a Corporeal Feminism*. Bloomington: Indiana University Press.
———. 2008. "Darwin and Feminism: Preliminary Investigations for a Possible Alliance." In *Material Feminisms*, ed. Stacy Alaimo and Susan Hekman, 23–51. Bloomington: Indiana University Press.

———. 2011. *Becoming Undone: Darwinian Reflections on Life, Politics, and Art.* Durham, NC: Duke University Press.
———. 2017. *The Incorporeal: Ontology, Ethics, and the Limits of Materialism.* New York: Columbia University Press.
Hacking, Ian. 1983. *Representing and Intervening: Introductory Topics in the Philosophy of Natural Science.* Cambridge: Cambridge University Press.
Hammonds, Evelynn, and Rebecca Herzig. 2009. Introduction to *The Nature of Difference: Sciences of Race in the United States from Jefferson to Genomics*, ed. Evelynn Hammonds and Rebecca Herzig, xi–xv. Cambridge, MA: MIT Press.
Hankinson Nelson, Lynn. 1990. *Who Knows: From Quine to a Feminist Empiricism*. Philadelphia: Temple University Press.
———. 2017. *Biology and Feminism: A Philosophical Introduction*. New York: Cambridge University Press.
Haraway, Donna. 1985. "A Manifesto for Cyborgs: Science, Technology, and Socialist Feminism in the 1980s," *Socialist Review* 15(2):65–107.
———. 1988. "Situated Knowledges: The Science Question in Feminism and the Privilege of Partial Perspective." *Feminist Studies* 14(3):575–99.
———. 1991. *Simians, Cyborgs, and Women: The Reinvention of Nature*. New York: Routledge.
———. 1997. *Modest_Witness@Second_Millenium.FemaleMan©_Meets_Oncomouse™: Feminism and Technoscience*. New York: Routledge.
———. 2003. *Companion Species Manifesto: Dogs, People, and Significant Otherness.* Chicago: Prickly Paradigm Press.
———. 2004. *Crystals, Fabrics and Fields: Metaphors That Shape Embryos*. Berkeley, CA: North Atlantic Books.
———. 2008a. "Otherworldly Conversations, Terran Topics, Local Terms." In *Material Feminisms*, ed. Stacy Alaimo and Susan Hekman, 157–87. Bloomington: Indiana University Press.
———. 2008b. *When Species Meet*. Minneapolis: University of Minnesota Press.
———. 2015. "Anthropocene, Capitalocene, Plantationocene, Chthulecene: Making Kin." *Environmental Humanities* 6:159–65.
———. 2016. *Staying with the Trouble: Making Kin in the Chthulucene*. Durham, NC: Duke University Press.
Harding, Sandra. 1986. *The Science Question in Feminism*. Ithaca, NY: Cornell University Press.
———. 1987. "Introduction: Is there a Feminist Method?" In *Feminism and Methodology: Social Science Issues*, ed. Sandra Harding, 1–14. Bloomington: Indiana University Press.
———. 1989. "Is There a Feminist Method?" In *Feminism and Science*, ed. Nancy Tuana, 17–32. Bloomington: Indiana University Press.
———. 1991. *Whose Science? Whose Knowledge? Thinking from Women's Lives.* Ithaca, NY: Cornell University Press.
———. 1993. "Introduction: Eurocentric Scientific Illiteracy—A Challenge for the World Community." In *The "Racial" Economy of Science: Toward a Demo-*

cratic Future, ed. Sandra Harding, 1–29. Bloomington: Indiana University Press.
———. 2008. *Sciences from Below: Feminisms, Postcolonialities, and Modernities*. Durham, NC: Duke University Press.
———. 2011. "Beyond Postcolonial Theory: Two Undertheorized Perspectives on Science and Technology." In *The Postcolonial Science and Technology Studies Reader*, ed. Sandra Harding, 1–31. Durham, NC: Duke University Press.
———. 2016. "Latin American Decolonial Social Studies of Scientific Knowledge: Alliances and Tensions." *Science, Technology, and Human Values* 41(6):1063–87.
Harding, Sandra, and Jean F. O'Barr, eds. 1987. *Sex and Scientific Inquiry*. Chicago: University of Chicago Press.
Hartmann, Martin. 2009. "Genital Mycoplasmas." *Journal of the German Society of Dermatology* 7:371–78.
Hartsock, Nancy, 1983. "The Feminist Standpoint: Developing the Ground for a Specifically Feminist Historical Materialism." In *Discovering Reality: Feminist Perspectives on Epistemology, Metaphysics, Methodology, and the Philosophy of Science*, ed. Sandra Harding and Merrill Hintikka, 283–310. Dordrecht: D. Reidel.
Hecht, Gabrielle. 2002. "Rupture Talk in the Nuclear Age: Conjugating Colonial Power in Africa." *Social Studies of Science* 32(5/6):691–727.
———. 2014. *Being Nuclear: Africans and the Global Uranium Trade*. Cambridge, MA: MIT Press.
Hekman, Susan. 2008. "Constructing the Ballast: An Ontology for Feminism." In *Material Feminisms*, ed. Stacy Alaimo and Susan Hekman, 85–119. Bloomington: Indiana University Press.
Hiltzik, Michael. 2017. "California's Stem Cell Program Ponders a Future of New Challenges and Old Promises." *Los Angeles Times*. February 3. www.latimes.com/business/hiltzik/la-fi-hiltzik-cirm-future-20170205-story.html.
Hird, Myra J. 2004. *Sex, Gender and Science*. New York: Palgrave Macmillan.
———. 2009. *The Origins of Sociable Life: Evolution after Science Studies*. New York: Palgrave Macmillan.
Holland, Eugene. 2013. *Deleuze and Guattari's A Thousand Plateaus*. New York: Bloomsbury.
hooks, bell. 1984. *Feminist Theory: From Margin to Center*. New York: South End Press.
Hubbard, Ruth. 1988. "Science, Facts, and Feminism." *Hypatia: A Journal of Feminist Philosophy* 3(1):5–17.
———. 1990. *The Politics of Women's Biology*. New Brunswick, NJ: Rutgers University Press.
———. 1995. "The Logos of Life." In *Reinventing Biology: Respect for Life and the Creation of Knowledge*, ed. Linda Birke and Ruth Hubbard, 89–102. Bloomington: Indiana University Press.

Human Microbiome Project. 2017. "The Human Microbiome." http://hmpdacc.org/overview/about.php.
Hutchison, Clyde A., Ray-Yuan Chuang, Vladimir N. Noskov, Nacyra Assad-Garcia, Thomas J. Deerinck, Mark H. Ellisman, John Gill, Krishna Kannan, Bogumil J. Kara, Li Ma, James F. Pellerier, Zhi-Qing Qi, R. Alexander Richter, Elizabeth A. Strychalski, Lijie Sun, Yo Suzuki, Billyana Tsvetanova, Kim S. Wise, Hamilton O. Smith, John I. Glass, Chuck Merryman, Daniel G. Gibson, and J. Craig Venter. 2016. "Design and Synthesis of a Minimal Bacterial Genome." *Science* 351(6280): aad6253.
Irigaray, Luce. 1985. *Speculum of the Other Woman.* Translated by Gillian C. Gill. Ithaca, NY: Cornell University Press.
Jagger, Gill. 2015. "The New Materialism and Sexual Difference." *Signs: Journal of Women in Culture and Society* 40(2):321–42.
Joel, Daphna. 2014. "Sex, Gender, and Brain: A Problem of Conceptualization." In *Gendered Neurocultures: Feminist and Queer Perspectives on Current Brain Discourses*, ed. Sigrid Schmitz and Grit Höppner, 169–86. Vienna: Zaglossus.
Joel, Daphna, Anelis Kaiser, Sarah Richardson, Stacey Ritz, Deboleena Roy, and Banu Subramaniam. 2015. "Lab Meeting: A Discussion on Experiments and Experimentation." *Catalyst: Feminism, Theory, Technoscience* 1(1):1–12.
Jordan-Young, Rebecca. 2010. *Brain Storm: The Flaws in the Science of Sex Differences.* Cambridge, MA: Harvard University Press.
Kaiser, Anelis, Sven Haller, Sigrid Schmitz, and Cordula Nitsch. 2009. "On Sex /Gender Related Similarities and Difference in fMRI Language Research." *Brain Research Reviews* 61:49–59.
Kakade, A. S., and Y. S. Kulkarni. 2014. "Mifepristone: Current Knowledge and Emerging Prospects." *Journal of the Indian Medical Association* 112(1):36–40.
Kay, Lily E. 1993. *The Molecular Vision of Life: Caltech, the Rockefeller Foundation, and the Rise of the New Biology*. New York: Oxford University Press.
———. 1996. "Life as Technology: Representing, Intervening, and Molecularizing." In *The Philosophy and History of Molecular Biology: New Perspectives*, ed. Sahotra Sarkar, 87–100. Boston: Kluwer Academic Publishers.
———. 2000. *Who Wrote the Book of Life? A History of the Genetic Code*. Stanford, CA: Stanford University Press.
Keane, Helen, and Marsha Rosengarten. 2002. "On the Biology of Sexed Subjects." *Australian Feminist Studies* 17(39):261–77.
Keasling, Jay. 2009. "Lecture: Synthetic Biology in Pursuit of Inexpensive, Effective, Anti-Malarial Drugs." *BioSocieties* 4:275–82.
Keller, Evelyn Fox. 1983. *A Feeling for the Organism: The Life and Work of Barbara McClintock*. New York: W. H. Freeman and Company.
———. 1985. *Reflections on Gender and Science*. New Haven, CT: Yale University Press.
———. 1989. "The Gender/Science System: Or, Is Sex to Gender As Nature Is to Science." In *Feminism and Science*, ed. Nancy Tuana, 33–44. Bloomington: Indiana University Press.

Keller, Evelyn Fox, and Helen E. Longino. 1996. Introduction to *Feminism and science*, ed. Evelyn Fox Keller and Helen Longino, 1–14. Oxford: Oxford University Press.

Kember, Sarah. 2003. *Cyberfeminism and Artificial Life*. New York: Routledge.

Kfoury, Charlotte. 2007. "Therapeutic Cloning: Promises and Issues." *McGill Journal of Medicine* 10(2):112–20.

Kirby, Vicky. 1997. *Telling Flesh: The Substance of the Corporeal*. New York: Routledge.

———. 2006. *Judith Butler: Live Theory*. London: Continuum International Publishing Group.

———. 2008. "Natural Convers(at)ions: Or, What If Culture Was Really Nature All Along?" In *Material Feminisms*, ed. Stacy Alaimo and Susan Hekman, 214–36. Bloomington: Indiana University Press.

———. 2011. *Quantum Anthropologies: Life at Large*. Durham, NC: Duke University Press.

———. 2012. "Initial Conditions." *differences: A Journal of Feminist Cultural Studies* 23(3):197–205.

Koutroufinis, Spyridon A. 2014. "Introduction: The Need for a New Biophilosophy." In *Life and Process: Towards a New Biophilosophy*, ed. Spyridon A. Koutroufinis, 1–36. Berlin: De Gruyter.

———. 2017. "Organism, Machine, Process. Towards a Process Ontology for Organismic Dynamics." *Organisms: Journal of Biological Sciences* 1(1):19–40.

Kuiper, George G. J. M., Eva Enmark, Markku Pelto-Huikko, Stefan Nilsson, and Jan-Åke Gustaffson. 1996. "Cloning of a Novel Estrogen Receptor Expressed in Rat Prostate and Ovary." *Proceedings of the National Academy of Sciences USA* 93:5925–30.

Kumar, Dinesh, and Yashbir Singh Shivay. 2008. *Definitional Glossary of Agricultural Terms*. Volume 2. New Delhi: I.K. International Publishing House.

Lambert, P. A. 2005. "Bacterial Resistance to Antibiotics: Modified Target Sites." *Advanced Drug Delivery Reviews* 57(10):1471–85.

Landecker, Hannah. 2007. *Culturing Life: How Cells Became Technologies*. Cambridge, MA: Harvard University Press.

Landschoot, Peter. 2018. "The Cool-Season Turfgrasses: Basic Structures, Growth and Development." Department of Plant Science, Penn State University. http://plantscience.psu.edu/research/centers/turf/extension/fact sheets/cool-season.

Lartigue, Carole, John I. Glass, Nina Alperovich, Rempert Pieper, Prashanth P. Parmar, Clyde A. Hutchison III, Hamilton O. Smith, and J. Craig Venter. 2007. "Genome Transplantation in Bacteria: Changing One Species to Another." *Science* 317(5838):632–38.

Lather, Patti. 2007. *Getting Lost: Feminist Efforts toward a Double(d) Science*. Albany: State University of New York Press.

Latour, Bruno. 1997. Foreword to *Power and Invention*. By Isabelle Stengers, vii–xx. London: University of Minnesota Press.

———. 2007. *Reassembling the Social: An Introduction to Actor-Network-Theory*. Oxford: Oxford University Press.

Lawrence, J. G. 1999. "Gene Transfer, Speciation, and the Evolution of Bacterial Genomes." *Current Opinion in Microbiology* 2(5):519–23.

Lee, Rachel. 2013. Introduction. *The Scholar and Feminist Online* 11(3).

———. 2014. *The Exquisite Corpse of Asian America: Biopolitics, Biosociality, and Posthuman Ecologies*. New York: New York University Press.

Le Guin, Ursula. 1976 [1969]. "Nine Lives." In *The Wind's Twelve Quarters*. New York: Bantam Books.

Lentzos, Fillippa, Gaymon Bennett, Jeff Boeke, Drew Endy, and Paul Rabinow. 2008. "Roundtable on Synthetic Biology: Visions and Challenges in Redesigning Life." *BioSocieties* 3:311–23.

Leung, Yuet-Kin, Paul Mak, Sazzad Hassan, and Shuk-Mei Ho. 2006. "Estrogen Receptor (ER)-β signaling." *Proceedings of the National Academies of Sciences USA* 103(35):13162–67.

Levine, Jon. 1997. "New Concepts of the Neuroendocrine Regulation of Gonadotropin Surges in Rats." *Biological Reproduction* 56:293–302.

Lewontin, Richard. C. 1991. *Biology as Ideology, the Doctrine of DNA*. Toronto: Anansi Press Limited.

Longino, Helen. 1989. "Can There Be a Feminist Science?" In *Feminism and Science*, ed. Nancy Tuana, 45–57. Bloomington: Indiana University Press.

———. 1990. *Science as Social Knowledge: Values and Objectivity in Scientific Inquiry*. Princeton, NJ: Princeton University Press.

———. 1993. "Subjects, Power and Knowledge: Description and Prescription in Feminist Philosophies of Science." In *Feminist Epistemologies*, ed. Linda Alcoff and Elizabeth Potter, 101–20. New York: Routledge.

Long Now Foundation. 2008. "Drew Endy, Jim Thomas: Synthetic Biology Debate." http://longnow.org/seminars/02008/nov/17/synthetic-biology-debate.

Lorde, Audre. 1984a. "Age, Race, Class and Sex: Women Redefining Difference." In *Sister Outsider: Essays and Speeches*, 114–23. Trumansburg, NY: Crossing Press.

———. 1984b. "Uses of the Erotic, the Erotic as Power." In *Sister Outsider: Essays and Speeches*, 53–59. Trumansburg: Crossing Press.

MacGillivray, Margaret, Akira Morishima, Felix Conte, Melvin Grumback, and Eric Smith. 1998. "Pediatric Endocrinology Update: An Overview. The Essential Roles of Estrogens in Pubertal Growth, Epiphyseal Fusion and Bone Turnover: Lessons from Mutations in the Genes for Aromatase and the Estrogen Receptor." *Hormone Research* 49(Suppl 1):2–8.

MacKenzie, Adrian. 2013. "Synthetic Biology and the Technicity of Biofuels." *Studies in History and Philosophy of Biological and Biomedical Sciences* 44:190–98.

Mahajan, Damodar K., and Steve N. London. 1997. "Mifepristone (RU486): A Review." *Fertility and Sterility* 68(6):967–76.

Margulis, Lynn. 1998. *Symbiotic Planet: A New Look at Evolution*. New York: Basic Books.

Margulis, Lynn, and Dorion Sagan. 1986. *Origins of Sex: Three Billion Years of Genetic Recombination*. New Haven, CT: Yale University Press.

———. 1991. *Mystery Dance: On the Evolution of Human Sexuality*. New York: Summit Books.

———. 1997. *What Is Sex?* New York: Simon and Schuster Inc.

———. 2002. *Acquiring Genomes: A Theory of the Origin of Species*. New York: Basic Books.

Martin, Emily. 1987. *The Woman in the Body: A Cultural Analysis of Reproduction*. Boston: Beacon Press.

Massumi, Brian. 1987. "Realer Than Real: The Simulacrum According to Deleuze and Guattari." *Copyright* 1:90–97.

McClintock, Barbara. 1950. "The Origin and Behavior of Mutable Loci in Maize." *Proceedings of the National Academy of Sciences USA* 36(6):344–55.

McNeil, Maureen. 2005. "Introduction: Postcolonial Technoscience." *Science as Culture* 14(2):105–12.

———. 2007. *Feminist Cultural Studies of Science and Technology*. New York: Routledge.

Meighoo, Sean. 2016. *The End of the West and Other Cautionary Tales*. New York: Columbia University Press.

Merchant, Carolyn. 1990. *The Death of Nature: Women, Ecology, and the Scientific Revolution*. New York: HarperOne.

Messing, Karen, and Donna Mergler. 1995. "'The Rat Couldn't Speak, But We Can': Inhumanity in Occupational Health Research." In *Reinventing Biology: Respect for Life and the Creation of Knowledge*, ed. Lynda Birke and Ruth Hubbard, 21–49. Bloomington: Indiana University Press.

Mills, Catherine. 2010. "Continental Philosophy and Bioethics." *Bioethical Inquiry* 7:145–48.

Minnite, Lorraine, and Frances Fox Piven. 2016. "Making Policy in the Streets." In *Urban Policy in the Time of Obama*, ed. James DeFilipis, 272–92. Minneapolis: University of Minnesota Press.

Mohanty, Chandra Talpade. 1984. "Under Western Eyes: Feminist Scholarship and Colonial Discourses." *boundary 2: an international journal of literature and culture* 12(3):333–58.

———. 2003. *Feminism without Borders: Decolonizing Theory, Practicing Solidarity*. Durham, NC: Duke University Press.

MSNBC Online. 2004. "California Give Go-Ahead to Stem-Cell Research." November 3. www.msnbc.msn.com/id/6384390.

Murphy, Michelle. 2009. "Gender and Sex." In *The Palgrave Dictionary of Transnational History: From the Mid-19th Century to the Present Day*, ed. Akira Iriye and Pierre-Yves Saunier, 438–41. London: Palgrave Macmillan.

———. 2012. *Seizing the Means of Reproduction: Entanglements of Feminism, Health, and Technoscience*. Durham, NC: Duke University Press.

———. 2013. "Distributed Reproduction, Chemical Violence, and Latency." *Scholar and Feminist Online* 11(3). http://sfonline.barnard.edu/life-un-ltd-feminism-bioscience-race.

Murugan, Varnee. 2009. "Embryonic Stem Cell Research: A Decade of Debate from Bush to Obama." *Yale Journal of Biology and Medicine* 82(3):101–103.

Myers, Natasha. 2015. *Rendering Life Molecular: Models, Modelers, and Excitable Matter*. Durham, NC: Duke University Press.

Nanda, Meera. 2003. *Prophets Facing Backwards: Postmodern Critiques of Science and Hindu Nationalism in India*. New Brunswick, NJ: Rutgers University Press.

Nandy, Ashis. 1995. *Alternative Sciences: Creativity and Authenticity in Two India Scientists*. Delhi: Oxford University Press.

National Institutes of Health. 2017. "NIH Policy and Guidelines on the Inclusion of Women and Minorities as Subjects in Clinical Research—Amended, Oct. 2001." https://grants.nih.gov/grants/funding/women_min/guidelines_amended_10_2001.htm.

National Science Foundation. 2014. "ADVANCE at a Glance." www.nsf.gov/crssprgm/advance/index.jsp.

NBC News. 2017. "Conservative Reps Urge Trump to Fire NIH Chief Francis Collins over Stem Cells." May 22. www.nbcnews.com/health/health-news/conservative-reps-urge-trump-fire-nih-chief-francis-collins-over-n763301.

New England Biolabs. 2018. "Molecular Cloning." Accessed March 9, 2018. www.neb-online.fr/neb/cloning-synthetic-biology.

NIH Stem Cell Information Home Page. 2016. "National Institutes of Health Guidelines for Human Stem Cell Research." Accessed March 10, 2018. https://stemcells.nih.gov/policy/2009-guidelines.htm.

NPR. 2006. "Pet-Cloning Business Closes Its Doors." October 15. www.npr.org/templates/story/story.php?storyId=6272448.

Ortiz, Fernando. 1947. *Cuban Counterpoint: Tobacco and Sugar*. New York: Alfred A. Knopf, Inc.

Oudshoorn, Nelly. 1994. *Beyond the Natural Body: An Archeology of Sex Hormones*. New York: Routledge.

Oyama, Susan, P. E. Griffiths, and R. D. Gray. 2003. Introduction to *Cycles of Contingency: Developmental Systems and Evolution*, ed. Susan Oyama, P. E. Griffiths, and R. D. Gray, 1–11. Cambridge, MA: MIT Press.

Paddon, C. J., P. J. Westfall, D. J. Pitera, K. Benjamin, K. Fisher, D. McPhee, M. D. Leavell, A. Tai, D. Eng, D. R. Polichuk, et al. 2013. "High-level Semi-synthetic Production of the Potent Antimalarial Artemisinin." *Nature* 496(7446):528–32.

Parisi, Luciana. 2004. *Abstract Sex: Philosophy, Bio-Technology and the Mutations of Desire*. New York: Continuum.

———. 2010. "Event and Evolution." *Southern Journal of Philosophy* 48 (Spindel Supplement):147–64.

Parr, Adrian. 2005. *The Deleuze Dictionary*. New York: Columbia University Press.

Penfield, Christopher. 2014. "Rethinking Philosophy and Theology with Deleuze: A New Cartography." *Notre Dame Philosophical Reviews*. March 12. http://ndpr.nd.edu/news/rethinking-philosophy-and-theology-with-deleuze-a-new-cartography.

Peplow, Mark. 2016. "Synthetic Malaria Drug Meets Market Resistance." *Nature* 530:389–90.

Philip, Kavita. 2004. *Civilizing Natures: Race, Resources, and Modernity in Colonial South India*. New Brunswick, NJ: Rutgers University Press.

Philip, Kavita, Lilly Irani, and Paul Dourish. 2012. "Postcolonial Computing: A Tactical Survey." *Science, Technology, and Human Values* 37(1):3–29.

Piercy, Marge. 1976. *Women on the Edge of Time*. New York: Fawcett Crest Books.

Pollock, Anne. 2012. *Medicating Race: Heart Disease and Durable Preoccupations with Difference*. Durham, NC: Duke University Press.

Pollock, Anne, and Banu Subramaniam 2016. "Resisting Power, Retooling Justice: Promises of Feminist Postcolonial Technosciences." *Science, Technology, and Human Values* 41(6):951–66.

Potter, Elizabeth. 2001. *Gender and Boyle's Law of Gases*. Bloomington: Indiana University Press.

Prasad, Amit. 2014. *Imperial Technoscience: Transnational Histories of MRI in the United States, Britain, and India*. Cambridge, MA: MIT Press.

Prossnitz, Eric R., Jeffrey Arterburn, and Larry Sklar. 2007. "GPR30: A G Protein-coupled Receptor for Estrogen." *Molecular and Cellular Endocrinology* 265–66:138–42.

Protevi, John. 2012. "Deleuze and Life." In *The Cambridge Companion to Deleuze*, ed. Daniel W. Smith, 239–64. Cambridge: Cambridge University Press.

Rabinow, Paul, and Gaymon Bennett. 2009. "Synthetic Biology: Ethical Ramifications 2009." *Systems and Synthetic Biology* 3:99–108.

Reardon, Jenny. 2005. *Race to the Finish: Identity and Governance in an Age of Genomics*. Princeton, NJ: Princeton University Press.

Rich, Adrienne. 1976. *Of Woman Born: Motherhood as Experience and Institution*. New York: W.W. Norton and Company.

Richardson, Sarah. 2013. *Sex Itself: The Search for Male and Female in the Human Genome*. Chicago: University of Chicago Press.

Richardson, Sarah M., Leslie A. Mitchell, Giovanni Stracquadanio, Kun Yang, Jessica S. Dymond, James E. DiCarlo, Dongwon Lee, Cheng Lai Victor Huang, Srinivasan Chandrasegara, Yizhi Cai, et al. 2017. "Design of a Synthetic Yeast Genome." *Science* (6329):1040–44.

Ritz, Stacey. 2016. "Complexities of Addressing Sex in Cell Culture Research." *Signs: Journal of Women in Culture and Society* 42(2):307–27.

Ritz, Stacey, David Antle, Julie Coté, Kathy Deroy, Nya Fraleigh, Karen Messing, Lise Parent, Joey St.-Pierre, Cathy Vaillancourt, and Donna Mergler. 2014. "First Steps for Integrating Sex and Gender Considerations into Basic Experimental Biomedical Research." *The FASEB [Federation of American Societies for Experimental Biology] Journal* 28:4–13.

Ro, Dae. K., Eric M. Paradise, Mario Ouellet, Karl J. Fisher, Karyn L. Newman, John M. Ndungu, Kimberly A. Ho, Rachel A. Eachus, Timothy S. Ham, James Kirby, et al. 2006. "Production of the Antimalarial Drug Precursor Artemisinic Acid in Engineered Yeast." *Nature* 440:740–43.

Roberts, Dorothy. 1998. *Killing the Black Body: Race, Reproduction, and the Meaning of Liberty*. New York: Vintage.

———. 2012. *Fatal Invention: How Science, Politics, and Big Business Re-create Race in the Twenty-first Century*. New York: New Press.

Robinson, Howard. 2016. "Dualism." In *Stanford Encyclopedia of Philosophy*, ed. Edward N. Zalta. https://plato.stanford.edu/archives/win2016/entries/dualism.

Rogers, Lesley. 1988. "Biology, the Popular Weapon: Sex Differences in Cognitive Function." In *Crossing Boundaries: Feminisms and the Critique of Knowledges*, ed. Barbara Caine, E. A. Grosz, and Marie de Lepervanche, 43–51. Sydney: Allen and Unwin.

———. 2001. *Sexing the Brain*. New York: Columbia University Press.

Roosth, Sophia. 2009. "Screaming Yeast: Sonocytology, Cytoplasmic Milieus, and Cellular Subjectivities." *Critical Inquiry* 35(2):332–50.

Roosth, Sophia, and Astrid Schrader. 2012. "Feminist Theory out of Science: Introduction." *differences: A Journal of Feminist Cultural Studies* 23(3):1–8.

Rose, Hilary. 1994. *Love, Power, and Knowledge: Toward a Feminist Transformation of the Sciences*. Bloomington: Indiana University Press.

Rose, Nikolas. 2007. *The Politics of Life Itself: Biomedicine, Power, and Subjectivity in the Twenty-first Century*. Princeton, NJ: Princeton University Press.

Rosser, Sue V. 1986. *Teaching Science and Health from a Feminist Perspective*. New York: Teachers College Press.

———. 1989. "Feminist Scholarship in the Sciences: Where Are We Now and When Can We Expect a Theoretical Breakthrough?" In *Feminism and Science*, ed. Nancy Tuana, 3–14. Bloomington: Indiana University Press.

———. 1990. *Female-Friendly Science: Applying Women's Studies Methods and Theories to Attract Students*. New York: Teachers College Press.

Rossiter, Margaret. 1982. *Women Scientists in America: Struggles and Strategies to 1940*. Baltimore, MD: Johns Hopkins University Press.

Rouse, Joseph. 1996. *Engaging Science: How To Understand Its Practices Philosophically*. Ithaca, NY: Cornell University Press.

———. 2002. *How Scientific Practices Matter: Reclaiming Philosophical Naturalism*. Chicago: University of Chicago Press.

Roy, Deboleena. 2004. "Feminist Theory in Science: Working Towards a Practical Transformation." *Hypatia: A Journal of Feminist Philosophy* 19(1):255–79.

———. 2007. "Somatic Matters: Becoming Molecular in Molecular Biology." *Rhizomes: Cultural Studies in Emerging Knowledge* 14(Summer). www.rhizomes.net/issue14/roy/roy.html.

———. 2008a. "Asking Different Questions: Feminist Practices for the Natural Sciences." *Hypatia: A Journal of Feminist Philosophy* 23(4):134–57.

———. 2008b. "Should Feminists Clone? And If So, How? Notes from an Implicated Modest Witness." *Australian Feminist Studies* 23(56):225–47.
———. 2011. "Feminist Approaches to Inquiry in the Natural Sciences: Practices for the Lab." In *The Handbook of Feminist Research: Theory and Praxis*, ed. Sharlene Hesse-Biber, 313–30. London: Sage Publications.
———. 2012. "Cosmopolitics and the Brain: The Co-Becoming of Practices in Feminism and Neuroscience." In *Neurofeminism: Issues at the Intersection of Feminist Theory and Cognitive Science*, ed. Robyn Bluhm, Anne Jaap Jacobson, and Heidi Maibom, 175–92. New York: Palgrave Macmillan.
———. 2014. "Developing a New Political Ecology: Neurosceince, Feminism, and the Case of the Estrogen Receptor." In *Gendered Neurocultures: Feminist and Queer Perspectives on Current Brain Discourses*, ed. Sigrid Schmitz and Grit Hoppner, 203–22. Vienna: Zaglossus Press.
Roy, Deboleena, Nadia L. Angelini, and Denise D. Belsham. 1999. "Estrogen Directly Represses Gonadotropin-releasing Hormone (GnRH) Gene Expression in Estrogen Receptor-α (ER-α)- and ERβ-expressing GT1-7 GnRH Neurons." *Endocrinology* 140:5045–53.
Roy, Deboleena, Nadia Angelini, Hiroki Fujieda, Greg M. Brown, and Denise D. Belsham. 2001. "Cyclical Regulation of GnRH Gene Expression in GT1-7 GnRH-secreting Neurons by Melatonin." *Endocrinology* 142:4711–20.
Roy, Deboleena, and Denise D. Belsham. 2002. "Melatonin Receptor Activation Regulates Gonadotropin-releasing Hormone (GnRH) Gene Expression and Secretion in GT1-7 GnRH Neurons: Signal Transduction Mechanisms." *Journal of Biological Chemistry* 277:251–58.
Roy, Deboleena, and Banu Subramaniam. 2016. "Matter in the Shadows: Feminist New Materialism and the Practices of Colonialism." In *Mattering: Feminism, Science and Materialism*, ed. Victoria Pitts-Taylor, 23–42. New York: New York University Press.
Rubin, David. 2012. "An Unnamed Blank That Craved a Name: A Genealogy of Intersex as Gender." *Signs: Journal of Women in Culture and Society* 37(4):883–908.
Russ, Joanna. 1975. *The Female Man*. New York: Bantam Books.
Russell, John. 2004. "Venter Making Waves." *Bio-IT World*. April 15. www.bio-itworld.com/archive/041604/venter-making-waves.
Sagan, Dorion, and Lynn Margulis. 1988. *Garden of Microbial Delights: A Practical Guide to the Subvisible World*. New York: Harcourt Brace Jovanovich.
Sagan, Lynn. 1967. "On the Origin of Mitosing Cells." *Journal of Theoretical Biology* 14(3):255–74.
Said, Edward. 1979. *Orientalism*. New York: Vintage Books.
Salleh, Anna. 2007. "Stem-cell Research 'Boosting' Women's Egg Trade." ABC News Science Online. July 5. www.abc.net.au/news/2007-07-06/stem-cell-research-boosting-womens-egg-trade/91272.
Sandoval, Celeste M., Marites Ayson, Nathan Moss, Bonny Lieu, Peter Jackson, Sara P. Gaucher, Tizita Horning, Robert H. Dahl, Judith R. Denery, Derek A.

Abbott, Adam L. Meadows. 2014. "Use of Pantothenate As a Metabolic Switch Increases the Genetic Stability of Farnesene Producing *Saccharomyces cerevisiae*." *Metabolic Engineering* 25:215–26.

Sandoval, Chela. 2000. *Methodology of the Oppressed*. Minneapolis: University of Minnesota Press.

———. 2004. "U.S. Third World Feminism: The Theory and Method of Differential Oppositional Consciousness." In *The Feminist Standpoint Theory Reader*, ed. Sandra Harding, 195–210. New York: Routledge.

Sargent, Pamela. 1976. Introduction to *Bio-Futures: Science Fiction Stories about Biological Metamorphosis*, ed. Pamela Sargent. New York: Vintage Books.

———. 1978. *Cloned Lives*. New York: Fawcett Press.

Sarkar, Sahotra. 1996a. "Biological Information: A Skeptical Look at Some Central Dogmas of Molecular Biology." In *The Philosophy and History of Molecular Biology: New Perspectives*, ed. Sahotra Sarkar, 187–231. Boston: Kluwer Academic Publishers.

———. 1996b. Introduction to *The Philosophy and History of Molecular Biology: New Perspectives*, ed. Sahotra Sarkar, 1–13. Boston: Kluwer Academic Publishers.

———. 2005. *Molecular Models of Life: Philosophical Papers on Molecular Biology*. Cambridge, MA: MIT Press.

Schiebinger, Londa. 1991. *The Mind Has No Sex? Women in the Origins of Modern Science*. Cambridge, MA: Harvard University Press.

———. 1994. *Nature's Body: Gender in the Making of Modern Science*. New Brunswick, NJ: Rutgers University Press.

———. 1999. *Has Feminism Changed Science?* Cambridge, MA: Harvard University Press.

Sen Gupta, D. P. 2009. "Jagadish Chandra Bose: The Man and His Time." In *Remembering Sir J.C. Bose*, ed. D. P. Sen Gupta, M. H. Engineer, and V. A. Shepherd, 1–62. London: World Scientific Publishing Company.

Sen Gupta, D. P., M. H. Engineer, and V. A. Shepherd. 2009. Editors Note (back cover). *Remembering Sir J.C. Bose*, ed. D. P. Sen Gupta, M. H. Engineer, and V. A. Shepherd. London: World Scientific Publishing Company.

Seth, Suman. 2009. "Putting Knowledge in its Place: Science, Colonialism, and the Postcolonial." *Postcolonial Studies* 12(4):373–83.

Shaviro, Steven. 2005. "Cosmopolitics." *The Pinocchio Theory*. Accessed February 1, 2017. www.shaviro.com/Blog/?p=401.

Shevde, Nirupama, Amy Bendixen, Krista Dienger, and J. Wesley Pike. 2000. "Estrogens Suppress RANK Ligand-induced Osteoclast Differentiation via a Stromal Cell Independent Mechanism Involving c-Jun Repression." *Proceedings of the National Academy of Sciences USA* 97(14):7820–34.

Shildrick, Margrit. 2004. "Genetics, Normativity, and Ethics: Some Bioethical Concerns." *Feminist Theory* 5(2):149–65.

Shiva, Vandana. 1995. "Democratizing Biology: Reinventing Biology from a Feminist, Ecological, and Third World Perspective." In *Reinventing Biology:*

Respect for Life and the Creation of Knowledge*, ed. Lynda Birke and Ruth Hubbard, 50–71. Bloomington: Indiana University Press.

———. 1997. *Biopiracy: The Plunder of Nature and Knowledge*. Toronto: Between the Lines.

Shivers, B. D., R. E. Harlan, J. I. Morrell, and D. W. Pfaff. 1983. "Absence of Oestradiol Concentration in Cell Nuclei of LHRH-Immunoreactive Neurons. *Nature* 304:345–47.

Shou, Wenying, Sri Ram, and M. G. Vilar. 2007. "Synthetic Cooperation in Engineered Yeast Populations." *Proceedings of the National Academy of Sciences USA* 104(6):1877–82.

Smith, Daniel W. 2003. "Deleuze and Derrida, Immanence and Transcendence: Two Directions in Recent French Thought." In *Between Deleuze and Derrida*, ed. Paul Patton and John Protevi, 46–66. New York: Continuum.

Smith, Dorothy. 1974. "Women's Perspective as a Radical Critique of Sociology." *Sociological Inquiry* 44:7–13.

Smith, Linda Tuhiwai. 1999. *Decolonizing Methodologies: Research and Indigenous Peoples*. London: Zed Books.

Spanier, Bonnie. 1995. *Im/Partial Science: Gender Ideology in Molecular Biology*. Bloomington: Indiana University Press.

Spivak, Gayatri Chakravorty. 1985. "Three Women's Texts and a Critique of Imperialism." *Critical Inquiry* 12(1):243–61.

———. 1988a. "Can the Subaltern Speak?" In *Marxism and the Interpretation of Culture*, eds. Cary Nelson and Lawrence Grossberg, 271–313. Chicago: University of Illinois Press.

———. 1988b. "Subaltern Studies: Deconstructing Historiography." In *Selected Subaltern Studies*, ed. Ranajit Guha and Gayatri Spivak, 3–32. Oxford: Oxford University Press.

———. 1999. *A Critique of Postcolonial Reason: Toward a History of the Vanishing Present*. Cambridge, MA: Harvard University Press.

Star, Susan L. 1989. *Regions of the Mind: Brain Research and the Quest for Scientific Certainty*. Stanford, CA: Stanford University Press.

Stengers, Isabelle. 1997. *Power and Invention: Situating Science*. London: University of Minnesota Press.

———. 2000a. "Another Look: Relearning to Laugh." *Hypatia: A Journal of Feminist Philosophy* 15(4):41–54.

———. 2000b. *The Invention of Modern Science*. Minneapolis: University of Minnesota Press.

———. 2005. "Introductory Notes on an Ecology of Practices." *Cultural Studies Review* 11(1):183–96.

———. 2010. *Cosmopolitics I*. Minneapolis: University of Minnesota Press.

Styhre, Alexander. 2001. "The Nomadic Organization: The Postmodern Organization of Becoming." *Tamara: Journal for Critical Organization Inquiry* 1(4):1–12.

Subramaniam, Banu. 2000a. "Archaic Modernities: Science, Secularism, and Religion in Modern India." *Social Text* 64(18)3:67–86.

———2000b. "Snow Brown and the Seven Detergents: A Metanarrative on Science and the Scientific Method." *Women's Studies Quarterly* 28(1/2): 296–304.

———. 2014. *Ghost Stories for Darwin: The Science of Variation and the Politics of Diversity*. Chicago: University of Illinois Press.

Subramaniam, Banu, Laura Foster, Sandra Harding, Deboleena Roy, and Kim TallBear. 2017. "Feminism, Postcolonialism, Technoscience." In *The Handbook of Science and Technology Studies*, fourth edition, ed. Ulrike Felt, Ravyon Fouché, Clark Miller, and Laurel Smith-Doerr, 407–33. Cambridge, MA: MIT Press.

Subramaniam, Banu, and Angie Willey. 2017. "Introduction to Science Out of Feminist Theory Part One: Feminism's Sciences." *Catalyst: Feminism, Theory, Technoscience* 3(1):1–23. www.catalystjournal.org. ISSN: 2380–3312.

Suchman, Lucy. 1987. *Plans and Situated Actions: The Problem of Human/Machine Communication*. New York: Cambridge University Press.

Sunder Rajan, Kaushik. 2006. *Biocapital: The Constitution of Postgenomic Life*. Durham, NC: Duke University Press.

———. 2017. *Pharmocracy: Trials of Global Biomedicine*. Durham, NC: Duke University Press.

Sweet, Richard L. 2009. "Treatment Strategies for Pelvic Inflammatory Disease." *Expert Opinion on Pharmacotherapy* 10(5):823–37.

Synthetic Genomics. 2017. "ExxonMobil and Synthetic Genomics Report Breakthrough in Algae Biofuel Research." www.syntheticgenomics.com/exxon mobil-and-synthetic-genomics-report-breakthrough-in-algae-biofuel -research.

Tagore, Rabindranath. 1915. *The Gardener*. New York: Macmillan Company.

———. 1921 [1916]. *Home and the World*. London: Macmillan and Company Ltd.

———. 1960. *Gitabitan*. Kolkata: Visva Bharati.

TallBear, Kim. 2013. *Native American DNA: Tribal Belonging and the False Promise of Genetic Science*. Minneapolis: University of Minnesota Press.

Tavares, L. S., C. S. Silva, V. C. de Souza, V. L. da Silva, C. G. Diniz, and M. O. Santoz. 2013. "Strategies and Molecular Tools To Fight Antimicrobial Resistance: Resistome, Transcriptome, and Antimicrobial Peptides." *Frontiers in Microbiology* 4(412):1–11.

Terasawa, Ei, and Brian P. Kenealy. 2012. "Neuroestrogen, Rapid Action of Estradiol, and GnRH Neurons." *Frontiers in Neuroendocrinology* 22:364–75.

Thacker, Eugene. 2010. *After Life*. Chicago: University of Chicago Press.

ThermoFisher Scientific. 2017. "User Guide: pCR™8/GW/TOPO® TA Cloning Kit." Accessed April 5, 2017. www.thermofisher.com/order/catalog/product /K250020.

Thompson, Charis. 1999. "Confessions of a Bioterrorist: Subject Position and the Valuing of Reproductions." In *Playing Dolly: Technological Formations,*

Fantasies, and Fictions of Assisted Reproduction, ed. E. Ann Kaplan and Susan Squier, 189–219. New Brunswick, NJ: Rutgers University Press.

Times of India. 2016. "Rajasthan Government Won't Drop the Two-Child Policy: Kataria." April 2. http://timesofindia.indiatimes.com/city/jaipur/Rajasthan-govt-wont-drop-the-two-child-policy-Kataria/articleshow/51656767.cms.

Trinh, Minh-ha T. 1991. *When the Moon Waxes Red: Representation, Gender and Cultural Politics*. New York: Routledge.

Tuana, Nancy. 1989a. Preface. In *Feminism and Science*, ed. Nancy Tuana, vii–xi. Bloomington: Indiana University Press.

———. 1989b. "The Weaker Seed: The Sexist Bias of Reproductive Theory." In *Feminism and Science*, ed. Nancy Tuana, 147–71. Bloomington: Indiana University Press.

———. 2008. "Viscous Porosity: Witnessing Katrina." In *Material Feminisms*, ed. Stacy Alaimo and Susan Hekman, 188–213. Bloomington: Indiana University Press.

US Food and Drug Administration. 2013. "FDA Drug Safety Communication: Risk of Next-Morning Impairment after Use of Insomnia Drugs; FDA Requires Lower Recommended Doses for Certain Drugs Containing Zolpidem (Ambien, Ambien CR, Edluar and Zolpimist." www.fda.gov/drugs/drugsafety/ucm334033.html.

USA Today. 2005. "Frist, Bush Split on Stem Cell Stance." *USA Today*. August 31. https://usatoday30.usatoday.com/news/washington/2005-07-31-frist-stem-cells_x.htm.

———. 2007. "Court Upholds Calif. Stem Cell Agency." *USA Today*. February 26. https://usatoday30.usatoday.com/tech/science/2007-02-26-stem-cell-ruling_x.htm.

van Anders, Sari M. 2013. "Beyond Masculinity: Testosterone, Gender/Sex, and Human Social Behavior in a Comparative Context." *Frontiers in Neuroendocrinology* 34:198–210.

van der Tuin, Iris. 2008. "Deflationary Logic: Response to Sara Ahmed's Imaginary Prohibitions: Some Preliminary Remarks on the Founding Gestures of the 'New Materialism.'" *European Journal of Women's Studies* 15(4):411–16.

———. 2011. "New Feminist Materialisms." *Women's Studies International Forum* 34:271–77.

———. 2015. *Generational Feminism: New Materialist Introduction to a Generative Approach*. New York: Lexington Books.

Venter, J. Craig. 2007. *A Life Decoded: My Genome, My Life*. New York: Penguin.

Venter, J. Craig, Karin Remington, John F. Heidelberg, Aaron L. Halpern, Doug Rusch, Johanthan A. Eisen, Dongying Wu, Karen E. Nelson, William Nelson, Derrick E. Fouts, et al. 2004. "Environmental Genome Shotgun Sequencing of the Sargasso Sea." *Science* 304(5667):66–74.

ViaGen. 2017. "Initiate Cloning." https://viagenpets.com/product/initiate-cloning.

Wajcman, Judy. 1991. *Feminism Confronts Technology*. Oxford: Polity Press.

Waldby, Catherine. 2002. "Stem Cells, Tissue Cultures, and the Production of Biovalue." *Health: An Interdisciplinary Journal for the Social Study of Health, Illness, and Medicine* 6(3):305–23.
Waldby, Catherine, and Katherine Carroll. 2012. "Egg Donation for Stem Cell Research: Ideas of Surplus and Deficit in Australian IVF Patients' and Reproductive Donors' Accounts." *Sociology of Health and Illness* 34(4):513–28.
Waters, Erica, and Maxine Watson. 2015. "Live Substrate Positively Affects Root Growth and Stolon Direction in the Woodland Strawberry, *Fragaria vesca*." *Frontiers in Plant Science* 6(814). doi: 10.3389/fpls.2015.00814.
Weber, Jutta. 2006. "From Science and Technology to Feminist Technoscience." In *Handbook of Gender and Women's Studies*, ed. Kathy Davis, Mary S. Evans, and Judith Lorber, 397–414. London: Sage Publications.
Welchman, Alistair. 1997. "Machinic Thinking." In *Deleuze and Philosophy: The Difference Engineer*, ed. Keith Ansell Pearson, 211–27. New York: Routledge.
Weinstein, Jami. 2010. "A Requiem to Sexual Difference: A Response to Luciana Parisi's "Event and Evolution." *Southern Journal of Philosophy* 48(Spindel Supplement):165–87.
Westfall, Patrick J., Douglas J. Pitera, Jacob R. Lenihan, Diana Eng, Frank X. Woolard, Rika Tenetin, Tizita Horning, Hiroko Tsuruta, David J. Melis, Andrew Owens, et al. 2012. "Production of Amorphadiene in Yeast, and Its Conversion to Dihydroartemisinic Acid, Precursor to the Antimalarial Agent Artemisinin." *Proceedings of the National Academy of Sciences* 109(3): E111–E118.
Wetsel, W. C., P. L. Mellon, R. I. Weiner, and A. Negro-Vilar. 1991. "Metabolism of Pro-luteinizing Hormone-releasing Hormone in Immortalized Hypothalamic Neurons." *Endocrinology* 129(3):1584–95.
Wilhelm, Kate. 1998 [1976]. *Where Late the Sweet Birds Sang*. New York: Orb Books.
Willey, Angie. 2016. *Undoing Monogamy: The Politics of Science and the Possibilities of Biology*. Durham, NC: Duke University Press.
Wilson, Elizabeth. 1998. *Neural Geographies: Feminism and the Microstructure of Cognition*. New York: Routledge.
———. 1999. "Introduction: Somatic Compliance—Feminism, Biology and Science." *Australian Feminist Studies* 14(29):7–18.
———. 2004. *Psychosomatic: Feminism and the Neurological Body*. Durham, NC: Duke University Press.
World Health Organization. 2015. *Guidelines for the Treatment of Malaria*. Third edition. Geneva: World Health Organization Press.
Wylie, Alison. 2002. *Thinking from Things: Essays in the Philosophy of Archeology*. Berkeley: University of California Press.
———. 2004. "Why Standpoint Matters." In *The Feminist Standpoint Theory Reader*, ed. Sandra Harding, 339–52. New York: Routledge.

Yen, Samuel S. C. 1991. “Hypothalamic Gonadotropin-releasing Hormone: Basic and Clinical Aspects. In *Brain Endocrinology*, ed. Marcella Motta, 245–80. New York: Raven Press.

Zou, Z., L. F. Horowitz, J. P. Montmayeur, S. Snapper, and L. B. Buck. 2001. “Genetic Tracing Reveals a Stereotyped Sensory Map in the Olfactory Cortex.” *Nature* 414:173–79.

———. 2008. “Retraction of Genetic Tracing Reveals a Stereotyped Sensory Map in the Olfactory Cortex.” *Nature* 452:120.

Index

Feminist Technosciences
Rebecca Herzig and Banu Subramaniam, *Series Editors*

Figuring the Population Bomb: Gender and Demography in the Mid-Twentieth Century, by Carole R. McCann

Risky Bodies & Techno-Intimacy: Reflections on Sexuality, Media, Science, Finance, by Geeta Patel

Reinventing Hoodia: Peoples, Plants, and Patents in South Africa, by Laura A. Foster

Queer Feminist Science Studies: A Reader, edited by Cyd Cipolla, Kristina Gupta, David A. Rubin, and Angela Willey

Gender before Birth: Sex Selection in a Transnational Context, by Rajani Bhatia

Molecular Feminisms: Biology, Becomings, and Life in the Lab, by Deboleena Roy

Lightning Source UK Ltd.
Milton Keynes UK
UKHW011558080320
359953UK00003B/738